FM:
Fundamentals
of
Mathematics
Volume I

Cecilia L. Cullen, *Editor*
Eileen M. Petruzillo, *Editor*

Jacob Cohen

Neal Ehrenberg

Alice Farkouh

Lorraine Largmann

Edward Williams

*The authors are supervisors of
mathematics in New York City public high schools.*

BARRON'S EDUCATIONAL SERIES, INC.

All inquiries should be addressed to:
Barron's Educational Series, Inc.
250 Wireless Boulevard
Hauppauge, New York 11788

Library of Congress Catalog Card No. 81-17538

Paper Edition
International Standard Book No. 0-8120-2501-6

Cloth Edition
International Standard Book No. 0-8120-5469-5

Library of Congress Cataloging in Publication Data
Main entry under title:

FM, fundamentals of mathematics.

   1. Mathematics—1961-   I. Cullen, Cecilia L.
QA39.2.F6   510   81-17538
ISBN 0-8120-2501-6 (v. 1) AACR2

PRINTED IN THE UNITED STATES OF AMERICA

23   510   13 12 11 10

Preface

To the Teacher Before your students can graduate, they will be expected to pass a minimum competency examination in mathematics. But what happens to the students who have entered high school without a solid foundation in basic math skills?

Barron's new *FM: Fundamentals of Mathematics* is a textbook designed to help the ninth-year prealgebra student assimilate basic skills while being introduced to essential topics in high school mathematics. Developed in conformity with the New York State Curriculum, it parallels the newly developed "Fundamentals of Mathematics" Curriculum used in New York City.

FM: Fundamentals of Mathematics is organized in two volumes, each of which can be taught over a term or a full year. The simply written, carefully planned lessons are supplemented by model problems, an abundance of exercises at graded levels of difficulty, and incisive chapter and vocabulary reviews. Three minimum competency practice exams are supplied.

The chapters are designed to help students develop mathematical reasoning skills, and each provides enough material for teachers to give cyclical homework assignments as well as classroom exercises. Answer keys will be available.

Volume 1 leads students to competency in the following topics: Formulas, Introduction to Algebra, Metric Measurement, Geometry, Managing Money, and Probability and Statistics.

To the Student Before you graduate, you are required to pass a minimum competency examination in mathematics. This textbook is designed to help you prepare for that test and to give you a foundation for algebra.

Read the Study Guide carefully. Refer to it in answering the questions in the Fact Finding. Study the Model Problem in each lesson and do the Exercises. You may be surprised at how many problems you know how to do. Try the Extra for Experts. You will become an expert if you practice.

You will succeed with time, patience, and practice. Try each problem. Skip the ones that you think are too difficult and return to them later. Remember that no question is unimportant. Do not hesitate to ask your teacher for assistance.

As you move ahead in the book, review the previous lessons by going back and doing examples that were not assigned by your teacher.

Use Measuring Your Vocabulary and Measuring Your Progress to prepare for your class tests. You will be pleased to see your test scores improve!

THE AUTHORS

January, 1982

Contents

Chapter 1 FORMULAS

Chapter 2 INTRODUCTION TO ALGEBRA

Chapter 3 METRIC MEASUREMENT

Chapter 4 GEOMETRY

Chapter 5 MANAGING MONEY

Chapter 6 PROBABILITY AND STATISTICS

FM:
Fundamentals
of
Mathematics
Volume I

Chapter 1
Formulas

1-1.1 *Writing Numerical Expressions*

Signs and symbols are part of daily life. You see a symbol and you understand its meaning. A red traffic light means ''stop.'' A traffic sign shaped like the one at the right also means ''stop.''

In arithmetic there are four ''*operations*,'' called *addition, subtraction, multiplication,* and *division*. These are called *bi*nary operations because they combine *two* numbers in a specific way to get one number as a result.

There are many words for the binary operations and several symbols to represent them. The table below lists the operations, examples of the way these operations are expressed in symbols, and some common English expressions indicating each operation.

Operation	Example	Key Words or Phrases
Addition	$7 + 6$ (+) *Result:* 13	*add* 7 and 6; the *sum* of 7 and 6; 6 *more than* 7; 7 *increased by* 6; the number that *exceeds* 7 *by* 6
Subtraction	$25 - 9$ (−) *Result:* 16	the *difference* between 25 and 9; 25 *minus* 9; 9 *less than* 25; *subtract* 9 *from* 25; 25 *decreased by* 9; 25 *diminished by* 9
Multiplication	4×6 (×) $4 \cdot 6$ (raised dot) 4(6) (parentheses (4)6 around one number, or (4)(6) both numbers, with no symbol between them) *Result:* 24	4 *times* 6; the *product of* 4 and 6; 4 and 6 are *factors* of the number 24
Division	$27 \div 9$ (÷) $9\overline{)27}$ ( ) ‾) $\dfrac{27}{9}$ (fraction line) *Result:* 3	27 *divided by* 9; the *quotient* of 27 and 9; the *ratio* of 27 to 9; 27 *over* 9

FACT FINDING

1. An operation that begins with two numbers and combines them in some way to result in one number is called a _____ operation.

2. The four binary operations of arithmetic are _____, _____, _____, and _____.

3. The product of 3 and 8 is 24. The numbers 3 and 8 are called _____ of the number 24.

4. Using the words "quotient," "difference," and "product," complete the following statements:

   a. The result of addition is called the sum. The result of subtraction is called the _____.

   b. The result of multiplication is called the _____.

   c. The result of division is called the _____.

MODEL PROBLEMS

1. Write a numerical expression for this English phrase:

   the sum of thirty-six and nineteen.

   Solution: "The sum of" indicates the operation *addition*.
   "A numerical expression" means write only numerals and operations.
   Answer: 36 + 19.

2. Write a numerical expression for this English phrase and evaluate:

   six hundred three decreased by eighty-seven.

   Solution: "Decreased by" indicates subtraction.
   Numerical expression: 603 − 87.
   Evaluation (compute):   603
                          − 87
                          ───────
                            516
   Answer: 603 − 87 = 516.

Complete the following table by providing all the missing information:

English Phrase	Operation Indicated	Numerical Expression	Evaluation
a. Five plus six	Addition	5 + 6	5 + 6 —— 11
b. The difference between eighteen and five	Subtraction	18 − 5	18 − 5 —— 13
c. Add five and nine			
d. Twelve added to seven			
e. Nine less than ten			
f. Five subtracted from forty-one			
g. Eight times seven			
h. The product of six and two			
i. Divide sixty by twelve			
j. Fifteen more than eight			
k. Nine multiplied by seventy			
l. Take fifteen from one thousand			
m. Seventeen less nine			
n. Forty-three exceeded by thirty-eight			
o. Eighty-four over seven			
p. Nine increased by six			
q. Twice twelve			
r. From thirty subtract seventeen			
s. From two hundred take ninety-three			
t. The quotient of forty divided by five			
u. Fifty diminished by thirty-two			
v. Fifty more than thirty-two			
w. The sum of twelve and eighty-nine			
x. Eighteen decreased by twelve			
y. The product of the factors nine and twelve			

1-1.2 *Writing Algebraic Expressions*

STUDY GUIDE Expressions containing only numerals and operations are called numerical expressions. $3 + 7$, $4 \cdot 5 - 2$, and $16 \div (8 - 3 \cdot 2)$ are three examples of numerical expressions. Most of the words and symbols used in writing numerical expressions are also used in algebra.

In algebra you often wish to describe *any number* rather than a specific number. In order to describe any number you use a symbol called a *variable* for the number. Variables are usually letters and they stand for numbers.

Expressions containing numerals, operations, and variables are called algebraic expressions. $3 + x$, $xy - a$, and $p \div (8 - 3m)$ are examples of algebraic expressions, and the letters x, y, a, p, and m are variables.

To show the product of 6 and the number p you can write:

$$6 \times p \qquad\qquad 6 \cdot p \qquad\qquad 6(p) \qquad\qquad 6p \qquad\qquad (6)(p).$$

The preferred way to show this multiplication is $6p$ with the number written in front of the variable and no operation sign between them.

FACT FINDING

1. Expressions containing only numerals and operations are called

_____ _____ .

2. A _____ is a symbol that stands for a number.

3. Expressions containing _____ , _____ , and

_____ are called algebraic expressions.

4. The preferred way to show the product of 9 and w is _____ .

MODEL PROBLEMS

1. Write in symbols an algebraic expression for the sum of ten and any number.

   Solution: Use the letter n to represent "any number."
   "The sum of" indicates addition.
   Answer: $10 + n$ or $n + 10$.

2. Express in symbols: h diminished by forty-seven.

   Solution: "Diminished by" indicates subtraction.
   Answer: $h - 47$.

1. Use the variable shown or choose your own variable to write each of the following as an algebraic expression:

    a. The sum of n and twelve   Solution: $n + 12$

    b. A number increased by ten   Solution: $x + 10$ (Use any variable you want.)

    c. Fourteen divided by a number

    d. x less twelve

    e. Nine times a number

    f. The difference between n and twelve

    g. Twelve subtracted from n

    h. The product of nine and p

    i. Fifteen diminished by y

    j. A number multiplied by fifteen

    k. A number added to three

    l. Some number decreased by twelve

    m. Some number less than twelve

    n. The quotient of a number divided by thirty

    o. Twenty more than w

    p. Twice x

    q. One half of a number

    r. A number over nine

    s. Fifteen increased by a number

    t. Twenty less than n

    u. Twenty decreased by n

    v. From twenty subtract n

    w. From n subtract fifteen

    x. The factors of a number are three and x

    y. Triple a number

2. None of the problems given below can be solved because there is a number missing. Write an algebraic expression to show the binary operation that would be used to solve the problem if the missing number were known.

    a. Bob rode the mechanical horse for 18 seconds. Juan rode the horse for n seconds more than Bob. For how many seconds did Juan ride the mechanical horse?   Solution: $18 + n$

    b. When going on a trip the Johnson family drove m miles on Monday and t miles on Tuesday. How many miles did they travel in all?

    c. The longest suspension bridge in the United States is the Verrazano Narrows Bridge, which is 1300 meters long. The Golden Gate Bridge is only x meters long. How many more meters in length is the Verrazano Bridge than the Golden Gate Bridge?

    d. There are 7 classes using this textbook. Each class has s students. How many students are using this textbook?

    e. The period 4 physical education class had y students. They were organized in 7 teams, each with the same number of players. How many students were placed on each team?

f. Sarah had a reading assignment of *r* pages. She read 12 pages as soon as she got home from school. How many pages did she have left to read?

g. The teacher assigned 24 questions to be answered but allowed the students to work as a group. If there were *g* students in the group, and each completed the same number of questions to share his or her results with the other members of the group, how many questions did each student complete on his or her own?

h. Francine has a part-time job that pays *x* dollars an hour. How much money will she earn in 25 hours?

i. A truck weighs *w* kilograms when it is empty. The truck is loaded with 2200 kilograms of cargo. What is the combined weight of truck and cargo?

j. There were 25 questions on a test. Zena answered *z* questions correctly. How many questions did she answer incorrectly?

1-2 *Evaluating Algebraic Expressions*

STUDY GUIDE An expression like $14 \cdot 3 - 2$ or $37 + 19 - 7$ is called a numerical expression or numerical phrase because it contains only numerals and operations. Every numerical expression can be evaluated by performing the operations.

Expressions containing variables, such as $12 + x$, are called *algebraic expressions*. Algebraic expressions take on values only when the variables are given specific numbers. For example, $12 + x$ becomes equal to 15 if the variable *x* is given the value 3. When specific values for the variable are *substituted* into the algebraic expression, a numerical expression is formed. In $12 + x$, the number 3 is put in place of *x*, resulting in the numerical expression $12 + 3$. Thus $12 + 3$ is evaluated by performing the indicated operation.

FACT FINDING

1. Expressions containing only numerals and operations are called

_____ _____ or _____ _____ .

2. _____ expressions contain variables.

3. When specific numbers are _____ for the variables in an algebraic expression, a _____ expression is formed.

4. _____ expressions can be evaluated by performing the indicated operations.

MODEL PROBLEMS

1. Evaluate $472 - x$ when $x = 289$.

   Solution: (1) *Substitute* the value for the variable x into the algebraic expression:

   $$472 - x \text{ becomes } 472 - 289.$$

   (2) *Evaluate* the resulting numerical expression:

   $$\begin{array}{r} 472 \\ -\ 289 \\ \hline 183 \end{array}$$

   (3) Write the answer: $472 - x = 183$.

2. Find the value of the sum of b and 12 decreased by c when $b = 42$ and $c = 18$.

   Solution: (1) *Translate* the expression into algebraic form:

   $$b + 12 - c.$$

   (2) *Substitute* the values for b and c into the algebraic expression:

   $$b + 12 - c \text{ becomes } 42 + 12 - 18.$$

   (3) *Evaluate:*

   $$\begin{array}{r} 42 \\ +\ 12 \\ \hline 54 \end{array} \qquad \begin{array}{r} 54 \\ -\ 18 \\ \hline 36 \end{array}$$

   (4) Write the answer: The sum of b and 12 decreased by c is 36.

EXERCISES

1. Evaluate each algebraic expression when the variables take on the indicated values.

a. $x + 97$; $x = 43$.   Solution: $\begin{array}{r} 43 \\ +\ 97 \\ \hline 140 \end{array}$ Answer: $x + 97 = 140$

b. $a - 143$; $a = 281$

c. $73 - b$; $b = 56$

d. $c + 621$; $c = 45$

e. $97 + d$; $d = 201$

f. $15e$; $e = 7$

g. $9f$; $f = 41$

h. $37(g)$; $g = 22$

i. $\dfrac{h}{15}$; $h = 75$

j. $2226 \div k$; $k = 21$

k. $m - 518$ when $m = 942$

l. $n - 809$ when $n = 4526$

m. $p + q + r$ when $p = 27$, $q = 197$, $r = 996$

n. $s - t$; $s = 6041$, $t = 762$

o. $u - v + w$; $u = 256$, $v = 137$, $w = 94$

p. $\dfrac{x}{y}$ when $x = 1989$ and $y = 51$

2. Translate each English phrase into an algebraic expression.

a. *a* divided by *b*

b. *c* decreased by *d*

c. *e* less *f*

d. *g* subtracted from 17

e. *h* multiplied by *k*

f. Twice *m*

g. *n* added to *p*

h. The sum of *q* and *r*

i. The product of *s*, *t*, and *u*

j. The number whose factors are *x*, *y*, and *z*

k. The sum of *a* and *b* decreased by *c*

l. The quotient of *d* divided by *e*

m. 356 more than *f*

n. Three times the product of *g* and *h*

o. Seven added to the sum of *j* and *k*

p. *m* diminished by *n*

q. 74 subtracted from *p*

r. *r* times *r*

s. *s* more than *t*

t. *u* less *v*

3. Write each algebraic expression in words.

a. *bh*   Solution: *b times h*

b. $\dfrac{d}{t}$ Solution: *d divided by t*

c. $a + b$

d. $c - d$

e. *ef*

f. $2g$

g. $h + i - j$

h. $2r + 2s$

i. $73 - p$

j. *prt*

k. $\dfrac{10}{x}$

l. $v + 9t$

4. Find the value of each algebraic expression when $w = 8$, $x = 5$, $y = 2$, and $z = 1$.

a. $x + y + z$

b. $x - y + z$

c. $\dfrac{w}{y}$

d. $3x$

e. $4xy$

f. $w + 3y$

g. wx

h. $2w - 7$

i. $wxyz$

j. $15 - w - z$

k. $x \cdot x \cdot x$

l. $3y(y)$

m. $\dfrac{x}{z}$

n. $\dfrac{w + y}{x}$

o. $x \cdot x + x \cdot y$

5. For each of the following English phrases:

   a. Translate the phrase into an algebraic expression.
   b. Evaluate the expression, using this table of values.

j	k	m	n
5	8	20	24

   (1) The sum of m and n

   Solution: (a) Translation: $m + n$
   (b) Evaluation: $m = 20$,
   $n = 24$,
   $m + n = 20 + 24 = 44$
   The sum of m and n is 44.

   (2) k subtracted from m
   (3) Thirty divided by j
   (4) j and n are factors of the product.
   (5) Twice m
   (6) n decreased by m
   (7) The sum of j, k, and m
   (8) Nine less than m
   (9) j more than k
   (10) Four times n plus seven times j
   (11) Fifteen added to nine times k

6. Solve each of the following problems.

   a. Lucia weighed f pounds, but she has lost 7 pounds since she went on a diet.
   (1) Write an algebraic expression to represent her present weight.
   (2) How much does she weigh now if f was 132 pounds?

   b. José had $392 in his savings account. He made a deposit of d dollars.
   (1) Represent his present bank balance.
   (2) If $d = \$45$, what is his present balance?

   c. Frank is y years old, Helen is 4 years older than Frank, and Rosemarie is three times as old as Frank.
   (1) Represent the age of each girl.
   (2) If y is equal to 5, how old is each girl?

   d. Amy, Beth, Meg, and Jo won c dollars in the lottery. They divided the winnings equally.
   (1) Represent each girl's share.
   (2) How much does each girl get if c is $10,000?

   e. A dozen eggs cost n cents.
   (1) Write an algebraic expression for the cost of one egg.
   (2) What is the cost of one egg if n is 96 cents?

Write the algebraic expression that answers each question.

a. Soda costs *s* cents per can, and nuts cost *j* cents a jar. What is the total cost of a 6-pack of soda and 2 jars of nuts?

b. A packing case contains *x* bottles of ketchup. Each bottle contains *y* milliliters. What is the total number of milliliters of ketchup contained in the packing case?

c. John walks to school. Each day he walks past a luncheonette that is directly between his home and his school. The luncheonette is *k* blocks from his home, and the school is *s* blocks from his home. How many blocks is the luncheonette from his school?

d. The school team scored *m* points in the first round of a tournament. In the second round of the tournament they scored *n* points more than they did in the first round. What was the total number of points scored by the school team in both rounds of the tournament?

e. The Robertsons drove *m* kilometers on a trip and used a total of *p* liters of gasoline. What was the average number of kilometers traveled for each liter of gasoline used?

1-3.1 *The Order of Operations*

Sometimes people interpret problems differently. When you are asked to evaluate the numerical expression $5 + 3 \times 2$, you might perform the operations in two different orders.

Multiply first:
$$\begin{array}{r} 3 \\ \times 2 \\ \hline 6 \end{array}$$

Add first:
$$\begin{array}{r} 5 \\ + 3 \\ \hline 8 \end{array}$$

Then add:
$$\begin{array}{r} 5 \\ + 6 \\ \hline 11 \end{array}$$
← two different values →

Then multiply:
$$\begin{array}{r} 8 \\ \times 2 \\ \hline 16 \end{array}$$

It is important for everyone to agree on a *single* value for a numerical expression. A numerical expression may not have two different values. A set of rules for the *order of operations* has been agreed upon. The *only* value for the numerical expression $5 + 3 \times 2$ is *11*. The rules are as follows.

In a numerical expression containing two or more operations:
1. Working from left to right, perform all multiplications or divisions as you come to them.
2. Working from left to right, perform all additions or subtractions as you come to them.

FACT FINDING

1. Every numerical expression must have only _____ value.

2. If you perform the operations in different orders, you may get

_____ _____ _____ .

3. It is important to agree upon a set of rules called the _____

_____ _____ so that everyone performs the operations
in the same order.

4. The first operations performed are _____ and division.

5. After all multiplications and divisions are performed, you go back to

the left of the resulting phrase and perform _____ or

_____ , proceeding in order from _____ to

_____ .

MODEL PROBLEM

Evaluate each numerical expression:
a. $18 - 6 \div 2$.

Solution: (1) Working from left to right, perform all
multiplications or divisions:
Do: $6 \div 2 = 3$.
The expression $18 - 6 \div 2$
becomes $18 - 3$

(2) Perform any additions or subtractions:
$18 - 3 = 15$.

(3) Write the answer: $18 - 6 \div 2 = 15$.

b. $14 \div 7 + 3 \cdot 6 - 9$.

Solution: (1) Working from left to right, perform all
multiplications or divisions as you come to them.
Do: $14 \div 7 = 2$.
Then do: $3 \cdot 6 = 18$.
The expression $14 \div 7 + 3 \cdot 6 - 9$
becomes $2 + 18 - 9$

(2) Working from left to right, perform all remaining
additions or subtractions as you come to them:
Do: $2 + 18 = 20$.
Then do: $20 - 9 = 11$.

(3) Write the answer: $14 \div 7 + 3 \cdot 6 - 9 = 11$.

1. Copy each numerical expression, and number the arrows to show the order in which computation is completed.

a. $9 + 2 \times 5$ Solution: $9 + 2 \times 5$ because you multiply
     ↑   ↑   ↑   ↑   first, then you add.
                             2   1

b. $12 - 6 \div 3$ **f.** $12 + 7 - 8$ **j.** $4 \div 2 \times 2 \div 10$
      ↑   ↑   ↑   ↑   ↑   ↑   ↑

c. $12 \div 6 - 3$ **g.** $30 \div 5 \times 2$ **k.** $4 - 2 \times 2 + 6$
      ↑   ↑   ↑   ↑   ↑   ↑   ↑

d. $12 - 6 + 3$ **h.** $4 \times 2 + 2 \times 6$ **l.** $12 - 6 \div 3 \times 2$
      ↑   ↑   ↑   ↑   ↑   ↑   ↑   ↑

e. $12 \times 6 \div 3$ **i.** $4 + 2 \times 2 + 6$ **m.** $10 + 9 \times 2 \div 3$
      ↑   ↑   ↑   ↑   ↑   ↑   ↑   ↑

2. Evaluate each numerical expression.

a. $6 + 3 \times 7$ **j.** $10 + 10 \div 10$ **s.** $15 + 15 \div 3 \times 2$

b. $6 \times 5 - 4$ **k.** $72 \div 9 - 1$ **t.** $40 - 5 \times 8 + 9$

c. $8 + 6 \times 2$ **l.** $24 \div 6 \div 3$ **u.** $16 + 9 \times 0 + 4$

d. $12 - 7 + 5$ **m.** $9 \times 2 + 5 \times 2$ **v.** $43 - 0 \div 7 + 5$

e. $46 - 6 \times 7$ **n.** $9 \times 2 \times 5 + 2$ **w.** $9 \times 0 \times 12 + 1$

f. $26 + 15 \times 9$ **o.** $9 + 2 \times 5 + 2$ **x.** $6 \cdot 3 - 2 \cdot 4$

g. $\dfrac{12}{3} - 4$ **p.** $12 \div 2 + 4 \times 4$ **y.** $9 \cdot 7 - \dfrac{84}{12}$

h. $15 + 6 \div 3$ **q.** $\dfrac{48}{6} + 9 \div 3$ **z.** $4 + 6 \times 5 \div 3 - 9$

i. $9 + \dfrac{9}{3}$ **r.** $72 \div 8 \times 3 + 1$

EXTRA FOR EXPERTS

Evaluate each numerical expression.

a. $14.2 - 7 \times 0.9$ **d.** $\dfrac{3}{4} \times \dfrac{1}{2} + \dfrac{2}{3}$

b. $17.2 + 26.8 \div 4$ **e.** $8.2 \div 14 \times \dfrac{3}{4}$

c. $28 \div \dfrac{1}{2} + 9$ **f.** $12 - 10 \times 0.5 + 8 \times \dfrac{7}{8}$

1-3.2 *Using the Order of Operations to Evaluate Algebraic Expressions*

STUDY GUIDE   Algebraic expressions can be evaluated after specific numerical values replace the variables. For example, $12x + 7$ can be evaluated for $x = 5$ if 5 is substituted for the x.

Once the substitution is completed, a *numerical expression* is formed. Thus the algebraic expression $12x + 7$ becomes the numerical expression $12 \cdot 5 + 7$.

The numerical expression is evaluated using the order of operations. Therefore $12 \cdot 5 + 7$ is evaluated with the result 67, and the expression $12x + 7$ has the value 67 when $x = 5$.

FACT FINDING

1. Algebraic expressions can be evaluated only after the variables have been replaced by specific _____.

2. The process of replacing the variables with numbers forming a numerical expression is called _____.

3. A numerical expression is evaluated using a set of rules called the _____ of operations.

MODEL PROBLEM

Evaluate $36 - 5y$ when $y = 6$.

Solution: (1) *Substitute* the value of y into the algebraic expression:
$36 - 5y$ becomes $36 - 5 \cdot 6$.
(*Remember:* $5y$ means $5 \cdot y$.)

(2) *Evaluate* $36 - 5 \cdot 6$, following the rules for the order of operations:
  a. Do multiplications or divisions first:
$$5 \cdot 6 = 30,$$
so $36 - 5 \cdot 6$ becomes $36 - 30$.
  b. Do additions or subtractions:
$$36 - 30 = 6.$$

(3) Write the answer: $36 - 5y = 6$.

1. For each of the following, write a statement that indicates the names of the operations and the order in which the operations are performed:

   a. $a + 4b$ Solution: Multiply 4 times b, then add a.

   b. $5a + 3$

   c. $7a + 6b$

   d. $3b - 2a$

   e. $a + b - 2$

   f. $4a + b - 7$

   g. $ab + 12$

   h. $15 - \dfrac{a}{b}$

   i. $\dfrac{a}{5} + \dfrac{b}{3}$

   j. $6 + 2a - \dfrac{b}{4}$

   k. $49 - 6ab$

2. Evaluate each algebraic expression using this table of values:

w	x	y	z
0	12	4	3

   a. $x + 2y$

   b. $23 - 4z$

   c. $6 + 5y$

   d. $\dfrac{x}{2} - y$

   e. $5w + 2$

   f. $x + y - 2z$

   g. $x - y + z$

   h. $\dfrac{4xy}{z}$

   i. $3z - y$

   j. $y \cdot y - \dfrac{x}{z}$

   k. $x \cdot y \cdot z \cdot w$

   l. $4x - 12y + w$

   m. $5zy - 2x$

   n. $19z + 8 - 5x$

3. Find the value of each algebraic expression for the values given.

   a. $a + 5b$ when $a = 42$ and $b = 17$

   b. $\dfrac{15x}{y}$ when $x = 12$ and $y = 5$

   c. $6m - n$ when $m = 18$ and $n = 15$

   d. $7pq - 6p$ when $p = 8$ and $q = 2$

   e. $12 - \dfrac{3y}{24} + 5x$ when $y = 64$, $x = 50$

   f. $3j - 15 - 4k$ when $j = 35$, $k = 8$

   g. $4b + 5bc - ab$ when $a = 12$, $b = 5$, $c = 2$

   h. $3xy - 2xy + z$ when $x = 5$, $y = 9$, $z = 15$

   i. $4d + 4e + 4f$ when $d = 7$, $e = 9$, $f = 14$

   j. $5ab - 5bc + 5ac$ when $a = 6$, $b = 5$, $c = 4$

4. Construct a table showing the value of the expression when the variable is each of the numbers given.

   a. $y + 9$ when $y = 0, 2, 4, 6$ Solution:

y	$y + 9$
0	9
2	11
4	13
6	15

   b. $5p$ when $p = 1, 3, 5, 7$

   c. $2x + 1$ when $x = 1, 2, 5, 7$

   d. $3y - 1$ when $y = 12, 14, 16, 18$

   e. $45 - 2n$ when $n = 7, 8, 9, 10$

   f. $2 + \dfrac{p}{4}$ when $p = 8, 12, 16, 20$

   g. $\dfrac{9n}{5}$ when $n = 15, 20, 25, 30$

   h. $15 - 4p$ when $p = 0, 1, 2, 3$

   i. $6j - 19$ when $j = 4, 7, 12, 15$

EXTRA FOR EXPERTS

Write each English phrase as an algebraic expression. Then evaluate the expression when $a = 3$ and $b = 12$.

a. Seven more than twice b

b. The product of a and b increased by seven

c. Four times a subtracted from sixteen

d. The sum of twice a and three times b

e. Ten times b decreased by seven times a

1-4.1 *Using Parentheses as Grouping Symbols*

STUDY GUIDE When the numerical expression $5 + 3 \times 2$ was introduced, it was stated that two people might interpret its numerical value differently. The order of operations was established to prevent misunderstandings. It was stated that multiplication and division were performed before addition and subtraction.

However, there are expressions in which you wish to have addition and subtraction performed first. *Grouping symbols* are used to indicate a change in the order of operations.

The most common grouping symbols are *parentheses:* ( ). The expression *inside* the parentheses must be evaluated first. Thus $(5 + 3) \times 2 = 16$ because $(5 + 3)$ must be evaluated first, giving the value (8), and $(8) \times 2 = 16$.

FACT FINDING

1. _____ _____ are used to change the order in which the operations are performed.

2. The most common grouping symbols look like this: ( ), and are called _____.

3. Expressions _____ parentheses are evaluated first.

MODEL PROBLEMS

1. Find the value of $5 + 2 \cdot (3 + 6)$.

   Solution: (1) *Evaluate* the expression *inside* the parentheses first: $5 + 2 \cdot (3 + 6)$ becomes $5 + 2 \cdot (9)$.

   (2) *Evaluate* the resulting expression. (*Remember:* Use the order of operations.)
   a. Multiply or divide:
   $5 + 2 \cdot (9) = 5 + 18$ $[2 \cdot (9) = 18]$.
   b. Add or subtract:
   $5 + 18 = 23$.

   (3) Write the answer: $5 + 2 \cdot (3 + 6) = 23$.

2. Find the value of $(2x + 9) \div 3$ when $x = 6$.

   Solution: (1) *Substitute* the value of x in the algebraic expression: $(2x + 9) \div 3$ becomes $(2 \cdot 6 + 9) \div 3$.

   (2) *Evaluate* the expression *inside* the parentheses first. (*Remember:* Use the order of operations.) $(2 \cdot 6 + 9) = (21)$ because $2 \cdot 6 = 12$ is done first, then $12 + 9 = 21$.

   (3) *Evaluate* the resulting numerical expression: $(21) \div 3 = 7$.

   (4) Write the answer: $(2x + 9) \div 3 = 7$.

EXERCISES

1. Copy each numerical expression, and number the arrows to show the order in which the computation is performed.

   a. $20 - (17 - 9)$ Solution: $20 - (17 - 9)$ because
        ↑   ↑   ↑   ↑   operations inside
                                       2   1   parentheses are
                                                     performed first.

b. $18 - (9 + 6)$
  ↑   ↑

c. $(4 + 7) \times 2$
  ↑   ↑

d. $4 + 7 \times 2$
  ↑   ↑

e. $2 \times (3 + 4) \times 5$
  ↑   ↑   ↑

f. $2 \times (3 + 4 \times 5)$
  ↑   ↑   ↑

g. $2 \times 3 + 4 \times 5$
  ↑   ↑   ↑

h. $(2 \times 3 + 4) \times 5$
  ↑   ↑   ↑

i. $48 \div 2 \times 4$
  ↑   ↑

j. $48 \div (2 \times 4)$
  ↑   ↑

k. $(16 + 3) \times (12 - 5)$
  ↑   ↑   ↑

2. Evaluate each pair of numerical expressions, and state whether the values are the same or different.

a. (1) $15 + (6 + 2)$ (2) $15 + 6 + 2$

Solution: (1) $15 + (6 + 2)$ (2) $\underbrace{15 + 6}_{21} + 2$
          $15 + \quad (8)$ $\underbrace{21 \quad + 2}$
              23 23

Answer: $15 + (6 + 2)$ has the same value as $15 + 6 + 2$. Both are 23.

b. (1) $40 - 16 + 12$ (2) $40 - (16 + 12)$

c. (1) $(47 - 35) + 15$ (2) $47 - 35 + 15$

d. (1) $5 \times (15 + 3)$ (2) $5 \times 15 + 3$

e. (1) $17 + 9 \times 6$ (2) $(17 + 9) \times 6$

f. (1) $6 + 3 \times 4$ (2) $6 + (3 \times 4)$

g. (1) $(14 + 2) \times 3$ (2) $14 + (2 \times 3)$

h. (1) $48 \div (6 + 2)$ (2) $48 \div 6 + 2$

i. (1) $15 + 13 - 9 - 2$ (2) $15 + (13 - 9) - 2$

j. (1) $(15 + 13) - (9 - 2)$ (2) $(15 + 13 - 9) - 2$

3. Evaluate each numerical expression.

a. $6(5 + 2)$ **k.** $6(5 + 4) - 9$

b. $27 - (9 + 3)$ **l.** $6(5 + 4 - 9)$

c. $72 \div (12 - 3)$ **m.** $(2 + 6)(5 + 8)$

d. $(36 - 10) \div 13$ **n.** $2 + 6(5 + 8)$

e. $15 - (15 \div 3)$ **o.** $(2 + 6)5 + 8$

f. $(15 - 15) \div 3$ **p.** $(25 + 15) \div (5 + 3)$

g. $40 \div (9 + 11)$ **q.** $25 + (15 \div 5) + 3$

h. $(26 + 13) \cdot 5$ **r.** $36 \div (4 - 2) + 7$

i. $4(3 + 2) + 6$ **s.** $36 \div (4 - 2 + 7)$

j. $28 - (4 \cdot 3 - 2)$ **t.** $(36 \div 4) - (2 + 7)$

4. Evaluate each algebraic expression, using $w = 8$, $x = 10$, $y = 4$, and $z = 12$.

    a. $2(x - 5)$ **f.** $(25 - 2)w$

    b. $y(z - x)$ **g.** $2z + 4(x - y)$

    c. $2w + y$ **h.** $\dfrac{(w + x)}{6}$

    d. $2(w + y)$ **i.** $24 \div (z - 2y)$

    e. $25 - 2w$ **j.** $\dfrac{5(4x - 32)}{2}$

5. Write a numerical expression, using parentheses, that expresses each of the following.

    a. The sum of 9 and 15 is found, and the result is multiplied by 12.

    b. Twenty is to be subtracted from the difference between thirty and nine.

    c. Nineteen is subtracted from forty, and six is added to the difference.

    d. Twice the sum of 9 and 5 is added to the product of 6 and 2.

    e. The sum of eighteen and nine is subtracted from the sum of twenty-two and sixteen.

EXTRA FOR EXPERTS

Insert parentheses so that each numerical expression has the indicated value.

Example: $3 + 12 \div 3 + 1 = 6$

Solution: $(3 + 12) \div 3 + 1$ or $3 + 12 \div (3 + 1)$
   because $(15) \div 3 + 1$ $3 + 12 \div (4)$
                        $5 + 1$ $3 + \quad\; 3$
                           6 6

    a. $10 - 5 - 1 = 6$ **f.** $25 - 3 \times 4 + 1 = 10$

    b. $2 + 3 \times 4 + 5 = 20$ **g.** $25 - 3 \times 4 + 1 = 12$

    c. $2 + 3 \times 4 + 5 = 45$ **h.** $6 \times 6 - 5 \times 5 = 30$

    d. $2 + 3 \times 4 + 5 = 25$ **i.** $6 \times 6 - 5 \times 5 = 151$

    e. $7 + 5 \cdot 7 - 6 = 78$ **j.** $20 - 3 \times 4 - 2 - 1 = 11$

1-4.2 *The Distributive Property of Multiplication over Addition*

STUDY GUIDE Some problems can be solved in more than one way. Bob and Frank tried to find the total number of boxes of cereal that were in a store display.

F	F	F	F	F	F	F	K	K	K	K	K	K	K	K	K	K	K
F	F	F	F	F	F	F	K	K	K	K	K	K	K	K	K	K	K
F	F	F	F	F	F	F	K	K	K	K	K	K	K	K	K	K	K
F	F	F	F	F	F	F	K	K	K	K	K	K	K	K	K	K	K
F	F	F	F	F	F	F	K	K	K	K	K	K	K	K	K	K	K
F	F	F	F	F	F	F	K	K	K	K	K	K	K	K	K	K	K
F	F	F	F	F	F	F	K	K	K	K	K	K	K	K	K	K	K
F	F	F	F	F	F	F	K	K	K	K	K	K	K	K	K	K	K
F	F	F	F	F	F	F	K	K	K	K	K	K	K	K	K	K	K

9 rows of Flakes — 7 boxes in each row

9 rows of Krispys — 11 boxes in each row

Bob solved the problem like this:

    7   boxes of Flakes in each row
  × 9   rows of Flakes
   63   boxes of Flakes

   11   boxes of Krispys in each row
  × 9 rows of Krispys
   99   boxes of Krispys

   63   boxes of Flakes
 + 99   boxes of Krispys
  162   boxes in display

Bob used this numerical expression to solve the problem:
$(9 \times 7) + (9 \times 11)$

Frank solved the problem like this:

     7 boxes of Flakes in each row
 + 11   boxes of Krispys in each row
    18 boxes of cereal in each row
 × 9   rows of cereal in the display
   162 boxes of cereal in the display

Frank used this numerical expression to solve the problem:
$9 \times (7 + 11)$

The problem can be solved in both ways. The two numerical expressions have the same value. $9 \times (7 + 11) = (9 \times 7) + (9 \times 11)$ is an example of the *distributive property of multiplication over addition.*

In the expression $9 \times (7 + 11)$ the factor, 9, distributes itself over each number in the sum, 7 and 11.

FACT FINDING

1. The expression $6 \times (2 + 3) = 6 \times 2 + 6 \times 3$ is an example of the

_____ property of _____ over _____ .

2. In the expression $9 \times (7 + 11)$, the 9 _____ itself over the 7

and the 11, giving the equivalent expression _____ $\times 7 +$

_____ $\times 11.$

MODEL PROBLEMS

1. Show that $7 \cdot (8 + 5)$ is equal to $7 \cdot 8 + 7 \cdot 5$.

   Solution: $7 \cdot (8 + 5)$ $=$ $7 \cdot 8 + 7 \cdot 5 =$

   $\qquad$ $7 \cdot \quad (13)$ $=$ $\quad 56 + 35 =$

   $\qquad\qquad$ 91 $\qquad\qquad\qquad$ 91

   Both expressions have the value 91.

2. Use the distributive property of multiplication over addition to find the missing number:

   $$6 \cdot (3 + 9) = 6 \cdot 3 + \underline{\quad} \cdot 9.$$

   Solution: 6, because the 6 in $6 \cdot (3 + 9)$ must be distributed over the 3 and the 9.

3. a. Determine whether the statement is TRUE or FALSE.
   b. State whether the statement is an example of the distributive property of multiplication over addition:

   $$11 \cdot 3 + 11 \cdot 4 = 11 \cdot 3 + 4$$

   Solution: a. $11 \cdot 3 + 11 \cdot 4 =$ $\qquad$ $11 \cdot 3 + 4 =$

   $\qquad\qquad$ $33 + 44 =$ $\qquad\qquad$ $33 + 4 =$

   $\qquad\qquad\qquad$ 77 $\qquad\qquad\qquad$ 37

   Answer: The statement is FALSE because $11 \cdot 3 + 11 \cdot 4 = 77$ and $11 \cdot 3 + 4 = 37$.

   b. The statement is NOT an example of the distributive property of multiplication over addition.

EXERCISES

1. Show that each statement is true by evaluating the numerical expression on each side of the "=" symbol.

   a. $5(4 + 7) = 5 \cdot 4 + 5 \cdot 7$

   b. $(6 + 9) \cdot 15 = 6 \cdot 15 + 9 \cdot 15$

   c. $12 \times (6 + 4) = 12 \times 6 + 12 \times 4$

   d. $11 \times 5 + 11 \times 15 = 11 \times (5 + 15)$

   e. $29 \times 63 + 29 \times 37 = 29 \times 100$

2. Find the missing number or numbers in each expression.

   a. $14 \times (3 + \underline{\quad}) = 14 \times 3 + 14 \times 5$

   b. $36 \times (8 + 2) = 36 \times \underline{\quad}$

   c. $14 \times (8 + 12) = \underline{\quad} \times 8 + \underline{\quad} \times 12$

   d. $23 \times (9 + 4) = 23 \times \underline{\quad} + 23 \times \underline{\quad}$

   e. $37 \times 2 + 53 \times 2 = (37 + 53) \times \underline{\quad}$

   f. $3 \cdot 7 + 17 \cdot 7 = 20 \cdot \underline{\quad}$

g. $\frac{1}{2} \cdot (9 + 5) = \underline{} \cdot 9 + \underline{} \cdot 5$

h. $\underline{} \cdot 31 + \underline{} \cdot 69 = 18 \cdot 100$

i. $16 \cdot 5 + 16 \cdot \underline{} = 16 \cdot (5 + 12)$

j. $32 \cdot 5 + 32 \cdot \underline{} = 32 \cdot 9$

EXTRA FOR EXPERTS

a. Determine whether each of the following statements is TRUE or FALSE.

b. Indicate which statements are examples of the distributive property of multiplication over addition.

(1) $11 \cdot (5 + 7) = 11 \cdot 5 + 11 \cdot 7$

(2) $11 \cdot 5 + 7 = (11 \cdot 5) + (11 \cdot 7)$

(3) $(11 \cdot 5) + 7 = (5 \cdot 11) + 7$

(4) $9 \cdot 7 + 9 \cdot 3 = 9(7 + 3)$

(5) $15 \cdot 20 = 15 \cdot 13 + 15 \cdot 7$

1-5 *Using Exponents*

STUDY GUIDE A convenient way to represent repeated multiplication is to use exponents. The expression 4^5 is read as 4 to the 5th power and means $4 \cdot 4 \cdot 4 \cdot 4 \cdot 4 = 1024$.

In 4^5 the 4 is called the *base* and the 5 is called the *exponent*. The exponent determines the number of times the base is used as a factor.

An exponent of 2 is read as the square of the number or the number squared. An exponent of 3 is read as the cube of the number or the number cubed.

FACT FINDING

1. Exponents are used to show repeated _____.

2. In the expression 4^5, 5 is called the _____ and 4 is called the

_____ .

3. The expression x^2 is read as the square of x or x _____ .

4. In evaluating 5^3, 5 is written as a _____ three times.

1. Express $4 \cdot 4 \cdot 5 \cdot 5 \cdot 5 \cdot 5$, using exponents.

    Solution: 4 is used as a factor two times:

    $$4 \cdot 4 = 4^2; \text{ read: 4 squared.}$$

    5 is used as a factor four times:

    $$5 \cdot 5 \cdot 5 \cdot 5 = 5^4; \text{ read: 5 to the 4th power.}$$

    Answer: $4 \cdot 4 \cdot 5 \cdot 5 \cdot 5 \cdot 5 = 4^2 \cdot 5^4$; read: 4 squared times 5 to the 4th power.

2. Evaluate $6 \cdot 10^3$.

    Solution: In $6 \cdot 10^3$, (read: 6 times 10 cubed), the exponent (3) refers only to the base (10), not to the 6. 10 is used as a factor 3 times.
    Therefore: $6 \cdot 10^3 = 6 \cdot 10 \cdot 10 \cdot 10$
    Multiply: $6 \cdot 10 = 60$, $60 \cdot 10 = 600$, $600 \cdot 10 = 6000$.
    Answer: $6 \cdot 10^3 = 6000$.

EXERCISES

1. Write each expression in exponential form.

    a. $3 \times 3 \times 3$

    b. 15×15

    c. $9 \times 9 \times 9 \times 9 \times 9 \times 9$

    d. $12 \times 12 \times 12 \times 12$

    e. $2.3 \times 2.3 \times 2.3 \times 2.3 \times 2.3$

    f. $7 \times 7 \times 7 \times 7 \times 8 \times 8 \times 8$

    g. $6 \times 10 \times 10 \times 10 \times 10 \times 10$

    h. $4 \times 4 \times 4 \times 9$

    i. $2 \times 2 \times 2 \times 2 \times 10 \times 10 \times 10$

    j. $1 \times 1 \times 1 \times 1 \times 1 \times 1 \times 1 \times 1 \times 1$

2. For each expression:

    a. Name the base and the exponent of the base.

    b. Write the expression in words.

(1) 6^3	(5) 25^{10}	(9) $4 \cdot 3^2$
(2) 5^2	(6) 1^{23}	(10) $5 \cdot 10^4$
(3) 5^7	(7) 23^1	(11) $6^4 \cdot 5^2$
(4) 2^8	(8) $4^2 \cdot 3$	(12) $2 \cdot 3$

3. Write a numerical expression for each of the following:

    a. Six to the fifth power
        f. Eighteen squared

    b. Ten to the fourth power
        g. Twenty to the ninth power

    c. Nine cubed
        h. Ten to the hundredth power

    d. One to the tenth power
        i. Two times the third power of ten

    e. The fifth power of two
        j. Five squared times three cubed

4. Write each expression in expanded form and evaluate.

    a. $2^4 + 3^2$ Solution: Expanded form: $2 \cdot 2 \cdot 2 \cdot 2 + 3 \cdot 3$.
                                   Evaluate, using
                                   the order of
                                   operations: $2^4 + 3^2 = 25$
                                                 $16 \ + 9 \ = 25$

b. 5^3		**f.** 2^7		**j.** $8 \cdot 10^3$		**n.** $3^4 - 2^3$
c. 1^2		**g.** $4^2 \cdot 3$		**k.** $6^2 \cdot 3^3$		**o.** $6^2 + 8^2$
d. 1^4		**h.** 10^4		**l.** 4^1		**p.** $2 \cdot 3 + 4^2$
e. 1^{15}		**i.** $6 \cdot 10^2$		**m.** $2^2 + 3^2$		**q.** $(2 + 3)^2$

    r. $4(1 + 2)^2$
               v. $3 \times 10^2 + 2 \times 10^1 + 9$

    s. $(12 - 2 \cdot 4)^2$
          w. $5 \times 10^3 + 1 \times 10^2 + 8 \times 10^1 + 2$

    t. $2 \times 10^1 + 7$
             x. $2 \times 10^3 + 0 \times 10^2 + 0 \times 10^1 + 1$

    u. $9 + 10^2 + 8 + 10^1$
        y. $4 \times 10^4 + 2 \times 10^2 + 3$

5. Tell whether each statement is TRUE or FALSE. Write an explanation for your conclusion.

    a. $6^2 = 12$ Solution: FALSE. 6^2 means 6×6, which
                                         equals 36, not 12.

b. $5^2 = 10$	**e.** $5^1 = 5$	**h.** $3^4 + 3^2 = 3^2$	
c. $2^4 = 4^2$	**f.** $3^4 = 4^3$	**i.** $2^2 + 2^2 = 4^4$	
d. $2^3 = 3^2$	**g.** $6^2 + 8^2 = 10^2$	**j.** $3^1 \times 3^4 = 3^5$	

**EXTRA
FOR
EXPERTS**

Find the value of each variable that will make each statement true.

    a. $572 = a \cdot 10^2 + b \cdot 10^1 + 2$
            d. $m^5 = 1$

    b. $963 = 9 \cdot 10^a + 6 \cdot 10^b + c$
       e. $p^3 = 50 + 2 \cdot 7$

    c. $y^2 = 81$

1-6 Evaluating Algebraic Expressions with Exponents

STUDY GUIDE   An algebraic expression has a specific value when a number is substituted for the variable in the given expression.

If the algebraic expression contains exponents, it is generally easier to evaluate if it is first rewritten without exponents.

For example, in order to evaluate the expression $3p^4$ for $p = 2$, $3p^4$ would be written as $3 \cdot p \cdot p \cdot p \cdot p$. The expression $3 \cdot p \cdot p \cdot p \cdot p$ is called the *expanded form* of $3p^4$. The value of the variable, 2, is substituted for p in the expanded form $3 \cdot p \cdot p \cdot p \cdot p$, yielding $3 \cdot 2 \cdot 2 \cdot 2 \cdot 2$. This numerical expression is equal to 48; thus $3p^4 = 48$ when $p = 2$.

FACT FINDING

1. To evaluate an algebraic expression you must substitute specific numerical values for the _____ in the expression.

2. $4 \cdot x \cdot x$ is called the _____ form of the expression $4x^2$.

3. In evaluating an algebraic expression containing exponents, you should first write the expression in _____ _____ and then _____ the given numerical value for the variable.

MODEL PROBLEMS

1. Evaluate $5n^3$ when $n = 4$.

   Solution: (1) *Write $5n^3$ in expanded form:*
   $$5n^3 = 5 \cdot n \cdot n \cdot n.$$

   (2) *Substitute* the value of n ($n = 4$) into the expanded form of the expression:
   $$5 \cdot 4 \cdot 4 \cdot 4.$$

   (3) *Evaluate:* $5 \cdot 4 = 20$, $20 \cdot 4 = 80$, $80 \cdot 4 = 320$.

   (4) Write the answer: $5n^3 = 320$.

2. Evaluate $16t^2 + 32t$ when $t = 3$.

   Solution: (1) *Write $16t^2 + 32t$ (read: $16t$ squared plus $32t$) in expanded form:*
   $$16 \cdot t \cdot t + 32 \cdot t.$$

   (2) *Substitute* the value of t into the expanded form of the algebraic expression:
   $$16 \cdot 3 \cdot 3 + 32 \cdot 3.$$

   (3) *Evaluate (Remember:* Use the order of operations.):
   $16 \cdot 3 \cdot 3 = 144$.
   $32 \cdot 3 = 96$, so $16 \cdot 3 \cdot 3 + 32 \cdot 3 = 144 + 96$.
   Then: $144 + 96 = 240$.

   (4) Write the answer: $16t^2 + 32t = 240$.

1. Write each expression in expanded form.

   a. m^3
   b. $m^2 \cdot n^3$
   c. $m \cdot n^3$
   d. $(m \cdot n)^3$

   e. $6x^2$
   f. $(3a)^2$
   g. $2x^2 + 3y^2$
   h. $(x + y)^2$

   i. $5x^2y^3$
   j. $3xy^2$
   k. $2^3 \cdot xy^4$
   l. $3(xy)^2$

2. Write each expression using exponents.

   a. $a \cdot a \cdot a$
   b. $b \cdot b$
   c. c cubed
   d. d squared
   e. The sixth power of e

   f. f to the fourth power
   g. g is used as a factor 6 times.
   h. 5 times the cube of h
   i. $12 \cdot j \cdot j \cdot j \cdot j$
   j. The square of $6k$

3. Evaluate each algebraic expression using the indicated value of the variable(s).

   a. c^2: $c = 5$
   b. $2m^2$: $m = 7$
   c. $(2m)^2$: $m = 7$
   d. cx^2: $c = 3$, $x = 4$
   e. a^2b^2: $a = 1$, $b = 3$
   f. $4x^2y$: $x = 7$, $y = 9$
   g. $2a^2 - b^2$: $a = 5$, $b = 3$
   h. $(2a - b)^2$: $a = 5$, $b = 3$

   i. $2(a^2 - b^2)$: $a = 5$, $b = 3$
   j. $a(a - b)^2$: $a = 5$, $b = 3$
   k. $23 - x^2$: $x = 4$
   l. $(23 - x)^2$: $x = 4$
   m. $3x^4$: $x = 10$
   n. $\frac{t^2}{12}$: $t = 6$
   o. $\frac{g^2}{h^3}$: $g = 4$, $h = 2$

4. Using the values $a = 2$, $b = 3$, and $c = 4$, evaluate each expression.

   a. a^2
   b. c^3
   c. a^4
   d. $2b^2$

   e. ab^2
   f. a^2b
   g. b^2c^2
   h. $3b^2c^2$

   i. $2abc^2$
   j. $(ab)^2$
   k. $b^2 + c^2$
   l. $c^2 - a^2$

   m. $a^2 + 2b^2 - c^2$
   n. $c^2 + 7$
   o. $(b + c)^2$
   p. $(c^2 - b^2)^2$

   q. $\frac{a^4}{c^2}$
   r. $a^2 + 2a - 3$
   s. $(3c - a^3)^2$
   t. $(2b^2 - c^2)^3$

Determine whether the expressions in each question are equivalent.

Example: $2^3 \cdot 5^2 | 10^3$

Solution: $\quad 2^3 \cdot 5^2 = \quad | \quad 10^3 =$

$\qquad 8 \cdot 25 = \quad | \quad 10 \cdot 10 \cdot 10$

$\qquad \quad 200 \qquad \qquad 1000$

Answer: The two expressions are not equivalent.

a. $2^3 + 2^3 | 4^3$ **f.** $x^2 \cdot x^3 \cdot x^4 | x^9$

b. $4^2 \cdot 4^2 \cdot 4^2 | 12^2$ **g.** $y^4 \cdot y^6 | y^{24}$

c. $3^2 \cdot 3^2 \cdot 3^2 | 27^2$ **h.** $x^3 \cdot y^3 | (xy)^3$

d. $5^2 \cdot 5^2 \cdot 5^2 | 5^6$ **i.** $8^9 \div 8^3 | 8^3$

e. $5^2 + 6^2 | 11^2$ **j.** $x^{10} \div x^4 | x^6$

1-7 *Evaluating Formulas*

STUDY GUIDE   It is often necessary to express a relationship among several quantities. A businessperson determines the selling price of an item by adding the cost of the item, the expenses involved in selling the item, and the desired profit. A formula for the businessperson's selling price could be:

$$s = c + e + p.$$

A *formula* is the mathematical statement of a rule expressing a relationship among quantities, using variables, numerals, operations, and an equality symbol. In a formula the variables are frequently the first letters of the key words in the rule.

A formula is used to find one quantity when all other quantities in the formula are known. The quantity that is to be found by using the formula is called the *subject* of the formula. In the formula $s = c + e + p$, s is the subject of the formula.

The subject of the formula is found by substituting the known values for the variables in the formula and evaluating the resulting numerical expression.

FACT FINDING

1. A mathematical statement of equality expressing a rule is called a

_____ .

2. In a formula the variables are frequently the _____

_____ of the key words.

3. The _____ is the quantity that can be found by using the formula.

4. A formula can be used to find _____ quantity when all other quantities in the formula are _____ .

MODEL PROBLEM

The formula for finding the Celsius temperature when the Fahrenheit temperature is known is:

$$C = \frac{5 \cdot (F - 32)}{9}.$$

Find the Celsius temperature corresponding to a Fahrenheit temperature of 95°.

Solution: Find C, the subject of the formula, when $F = 95$.
    (1) *Write* the formula:

$$C = \frac{5(F - 32)}{9}.$$

    (2) *Substitute* the known value, $F = 95$, into the formula:

$$C = \frac{5(95 - 32)}{9}.$$

    (3) *Evaluate* the numerical expression, using the rules for parentheses and the order of operations.
      a.   Parentheses first: $(95 - 32) = (63)$.

$$C = \frac{5(63)}{9}.$$

      b.   Multiplications or divisions in order:

$$5 \cdot (63) = 315 \qquad 9\overline{)315}.$$ (35)

    (4) Write the answer: When $F = 95°$, $C = 35°$.

EXERCISES

1. Translate each of the following rules into a formula, using the letters given:

    a. The cost, c, of sending a package is equal to ten times the weight, w, plus twelve.   Solution: $c = 10w + 12$

    b. The selling price, s, is equal to the cost, c, plus the profit, p.

    c. The selling price, s, is the original price, p, less the discount allowed, d.

    d. The diameter, d, is twice the radius, r.

e. The area, A, is the product of the length, l, and the width, w.

f. The number of days, d, is seven times the number of weeks, w.

g. The number of inches, i, is found by multiplying the number of feet, f, by twelve.

h. The interest gained, i, is the product of the principal, p, the rate, r, and the time, t.

i. The time, t, required to complete a trip is the distance, d, divided by the average rate of speed, r.

j. The value of a number, n, is ten times the tens digit, t, plus the units digit, u.

k. The number of centimeters, c, is 2.54 times the number of inches, i.

2. Write an English statement for the rule expressed by each formula.

a. $s = c + p$, where s = selling price, c = cost, p = profit.
    Solution: The selling price is equal to the cost plus the profit.

b. $A = p + i$, where A = amount, p = principal, i = interest.

c. $i = A - p$, where A = amount, p = principal, i = interest.

d. $p = 3s$, where p = perimeter of an equilateral triangle, s = length of one side of the triangle.

e. $t = \dfrac{d}{r}$, where t = time of travel, d = distance traveled, r = average rate of speed.

f. $I = \dfrac{V}{R}$, where I = current, V = voltage, R = resistance.

g. $W = I^2R$, where W = power in watts, I = current, R = resistance.

h. $E = mc^2$, where E is the energy, m is the mass of the object, and c is the speed of light.

i. $C = \dfrac{5(F - 32)}{9}$, where C is the Celsius temperature and F is the Fahrenheit temperature.

j. $F = \dfrac{9C}{5} + 32$, where C is the Celsius temperature and F is the Fahrenheit temperature.

k. $V = \dfrac{4\pi r^3}{3}$, where V is the volume of a sphere, π is pi (pi is equal to approximately 3.142), and r is the radius.

3. Evaluate the subject of each formula, given the values indicated.

a. $d = rt$: r = 82 km/hr, t = 8 hr

b. $c = np$: n = 12 items, p = 9 cents/item

c. $a = p + i$: p = 1200, i = 280

d. $s = p - d$: $p = 360$, $d = 42$

e. $r = \dfrac{d}{t}$: $d = 825$, $t = 15$

f. $p = 2a + 2b$: $a = 19$, $b = 34$

g. $p = 4s$: $s = 37$

h. $A = \dfrac{bh}{2}$: $b = 12$, $h = 9$

i. $V = e^3$: $e = 7$

j. $p = 2(l + w)$: $l = 227$, $w = 436$

4. Solve the following problems, using the formula(s) given in each problem.

   a. A cube is a geometric solid that has all edges equal. One particular cube has an edge that measures 6 centimeters.
      (1) Find the volume of this cube ($V = e^3$).
      (2) Find the total area of all the surfaces of this cube ($S = 6e^2$).

   b. A baseball diamond is in the shape of a square and is 90 feet on each side.
      (1) How much area is enclosed by the baseball diamond ($A = s^2$)?
      (2) What is the perimeter, that is, the measure around the edge, of the baseball diamond ($p = 4s$)?

   c. A rectangular solid measures 12 units long by 8 units wide by 5 units deep.
      (1) Find the volume of the solid ($V = lwd$).
      (2) What is the total surface area of the rectangular solid $[S = 2(lw + ld + wd)]$?

   d. Frank Wilson wishes to surround his flower garden with a wire fence. If the garden is 12 meters long by 18 meters wide, how much fencing does he need $[p = 2(l + w)]$?

   e. An area rug must be rebound after it is cleaned. Binding is material that is placed completely around the edges of the rug. If the rug is 9 feet long by 12 feet wide how much binding is needed ($p = 2l + 2w$)?

   f. An airplane maintains an average rate of 400 kilometers per hour for a total of 6 hours. What is the total distance covered by the airplane ($d = rt$)?

   g. A triangle has a base of 12 centimeters and a height of 9 centimeters. Find the area of the triangle $\left(A = \dfrac{bh}{2}\right)$.

   h. The Fahrenheit temperature is 95°. Find the corresponding reading on the Celsius thermometer $\left[C = \dfrac{5(F - 32)}{9}\right]$.

i. The weather report stated that the Celsius temperature range for a particular day would be between 20° and 25°. Find the temperature range expressed in Fahrenheit degrees $\left(F = \dfrac{9C}{5} + 32 \right)$.

j. A swimming pool is to be filled with water. The pool is 15 meters long and 10 meters wide, and averages 2 meters deep. How many liters of water are needed to fill the pool if $V = 10^3 \cdot lwd$?

k. To find the horsepower of a gasoline engine the formula $H = \dfrac{3d^2 n}{20}$ is used, where d represents the diameter of the cylinder in centimeters and n represents the number of cylinders. Find the horsepower of a gasoline engine with 4 cylinders, each of which is 12 centimeters in diameter.

EXTRA FOR EXPERTS

Formulas are rules written in algebraic form. When you follow a formula, you are following a set of rules. Here is a set of rules used to find the day of the week for any date in the twentieth century (1900–1999). Use these rules to find the dates listed.

To find the day of the week for the . . . month, . . . day, 19 . . . :

let w = the last two digits of the year

$x = \dfrac{w}{4}$, *disregarding any remainder*

y = the day number

z = month code (see right)

d = the *remainder* in the following: $\dfrac{w + x + y + z}{7}$

Table for Month Number

January	1	(leap year)	0
February	4	(leap year)	3
March	4		
April	0		
May	2		
June	5		
July	0		
August	3		
September	6		
October	1		
November	4		
December	6		

If $d = 1$, the day is Sunday. If $d = 2$, the day is Monday, and so on. Thus, if $d = 6$, the day is Friday. If there is no remainder ($d = 0$), the day is Saturday.

Find the day of the week of:

a. January 1, 1999

b. November 22, 1963

c. December 25, 1983

d. Your birth

Measuring Your Progress

SECTION 1-1

1. Name the operation indicated by each expression.

   a. The difference between 12 and x

   b. The sum of 19 and 47

   c. $12m$

   d. The product of 17 and 26

   e. The quotient of 225 and 45

   f. Fifteen increased by y

   g. $\dfrac{x}{41}$

   h. Twice nineteen

   i. Fifteen less than p

   j. Nine diminished by two

2. Write each phrase in symbols.

   a. The product of 5 and w

   b. Fourteen subtracted from p

   c. From 263 subtract 147.

   d. Fifty more than sixty-one

   e. A number divided by 23

   f. The factors of a number are 3 and m.

   g. 27 increased by 92

SECTION 1-2

1. Evaluate each algebraic expression for the indicated value of the variables.

   a. $c - d$: $c = 47$, $d = 29$

   b. $x + y - z$: $x = 175$, $y = 253$, $z = 124$

   c. ab: $a = 43$, $b = 19$

   d. $\dfrac{c}{d}$ when $c = 48$ and $d = 18$

   e. $x \cdot x \cdot x$ if $x = 5$

2. Find the value of each expression if x is 12 and y is 9.

   a. Four less than x

   b. Twice y

   c. Fifteen added to the product of x and y

   d. x diminished by y

   e. Eighty-seven divided by x

3. Six cans of soda cost c cents.

    a. Write an algebraic expression for the cost of one can of soda.

    b. Find the cost of one can of soda if $c = 96$ cents.

SECTION 1-3

1. Find the value of each numerical expression.

    a. $8 + 3 \cdot 2$ **d.** $8 + 3 \cdot 2 + 6$

    b. $12 \div 4 \times 3$ **e.** $17 \cdot 5 - 4 \cdot 12$

    c. $8 \times 3 + 2 \times 6$

2. Evaluate each algebraic expression for the indicated value of the variable(s).

    a. $x + 2y$: $x = 3, y = 6$ **d.** $6ab - b \cdot b$: $a = 9, b = 7$

    b. $43 - 5z$: $z = 8$ **e.** $14 - \dfrac{c}{9}$: $c = 72$

    c. $\dfrac{12x}{y}$: $x = 7, y = 6$

SECTION 1-4

1. Find the value of each numerical expression.

    a. $45 - (12 + 15)$ **d.** $15 - (12 \div 2)$

    b. $(6 + 7)5$ **e.** $25 - (3 \cdot 2 + 5)$

    c. $84 \div (12 - 5)$

2. Evaluate each algebraic expression, using $a = 5$, $b = 6$, and $c = 2$.

    a. $4(a - c)$ **d.** $a(b + c)$

    b. $16 - 2b$ **e.** $24 \div (b - a)$

    c. $\dfrac{(2a + 3b)}{7}$

3. Using parentheses, write the numerical expression indicated by each of the following. Evaluate the expression.

    a. Four is multiplied by the sum of 7 and 12.

    b. The difference between 28 and 16 is divided by 3.

    c. The sum of 42 and 26 is subtracted from 205.

4. Determine which of the following statements represent examples of the distributive property of multiplication over addition:

    a. $5 \times 2 + 5 \times 7 = 5 \times (2 + 7)$

    b. $6 \cdot (3 + 9) = 6 \cdot 3 + 9$

    c. $4 \cdot 20 = 4 \cdot 12 + 4 \cdot 8$

SECTION 1-5

1. Write a numerical expression in exponential form for each of the following:

    a. $6 \times 6 \times 6 \times 6$

    b. $9 \times 9 \times 10 \times 10 \times 10$

    c. Twelve cubed

    d. Seven is used as a factor six times.

    e. Two times the sixth power of ten

2. Evaluate each expression.

    a. 5^2 **d.** 5×10^3

    b. 2^5 **e.** $(6 + 3)^2$

    c. $2^2 + 3^2$

SECTION 1-6

1. Write each expression using exponents.

    a. $a \cdot a \cdot a \cdot b \cdot b$

    b. f is used as a factor 9 times.

    c. 6 times the square of k

    d. The square of $6k$

2. Write in expanded form.

    a. a^3 **b.** $6y^2$ **c.** $4x^5y^3$

3. Find the value of each expression for the indicated value of the variable(s).

    a. y^4: $y = 5$ **d.** $(a + b)^2$: $a = 3$, $b = 4$

    b. x^2y^2: $x = 3$, $y = 2$ **e.** $\dfrac{c^3}{12}$: $c = 6$

    c. $a^3 + b^2$: $a = 3$, $b = 4$

Evaluate the subject of each formula for the values indicated.

a. $c = np$: $n = 4$, $p = 72$

b. $t = \dfrac{d}{r}$: $d = 275$, $r = 55$

c. $A = s^2$: $s = 9$

d. $V = e^3$: $e = 5$

e. $K = \dfrac{b \cdot h}{2}$: $b = 16$, $h = 23$

f. $C = \dfrac{5(F - 32)}{9}$: $F = 86$

g. $w = I^2 r$: $I = 4$, $r = 200$

h. $p = 2a + 2b$: $a = 7$, $b = 9$

Measuring Your Vocabulary

Column II contains the meanings or descriptions of the words or phrases in Column I, which are used in this chapter.

For each number from Column I, write the letter from Column II that corresponds to the best meaning or description of the word or phrase.

Column I	Column II
1. Base	a. Combines two numbers to get one number as a result.
2. Binary operation	
3. Cube of a number	b. The result of addition.
4. Difference	c. The result of subtraction.
5. Distributive property of multiplication over addition	d. The result of multiplication.
	e. The result of division.
	f. One of the numbers used to form a product.
6. Exponent	g. The second power of a number.
7. Factor	h. The rules for evaluating numerical expressions.
8. Order of operations	
9. Product	i. Contains only numerals and operations.
10. Quotient	j. A letter used to represent a number.
11. Square of a number	k. $4 + (7 + 3) = (4 + 7) + 3$.
12. Subject of a formula	l. $4 \times (7 + 3) = 4 \times 7 + 4 \times 3$.
	m. $4 \times (7 + 3) = (7 + 3) \times 4$.
	n. 12 in the expression 12^3.
	o. 7 in the expression 5^7.
	p. $y \cdot y \cdot y$.
	q. A in the statement $A = b \cdot h$.

Maintaining Computational Skills—Whole Numbers

1. Addition:

a.	692	b.	124	c.	817	d.	123	e.	97
	825		707		909		45		168
	928		165		618		7236		2941
	539		826		29		170		402
									765

f. $716 + 92 + 29 + 574 + 980 =$

g. $76 + 457 + 746 + 59 + 824 =$

h. $1456 + 375 + 913 + 567 + 2849 =$

i. $4013 + 518 + 63 + 492 + 4573 =$

j. Find the sum of 609, 5137, 9, 63, and 5472.

2. Subtraction:

a.	974	d.	614	g.	903	j.	800	m.	8040
	-241		-541		-81		-376		-2753
b.	871	e.	1229	h.	905	k.	5970	n.	5006
	-659		-841		-616		-382		-2341
c.	960	f.	705	i.	945	l.	7003	o.	20,100
	-417		-562		-509		-284		-873

p. From 9031 subtract 843.

q. Subtract 8031 from 9169.

r. Find 526 less than 800.

s. Subtract 31 from 2000.

t. From 19,004 subtract 9578.

3. Multiplication:

a.	38	c.	72	e.	7428	g.	476	i.	9630
	$\times 46$		$\times 68$		$\times 54$		$\times 804$		$\times 305$
b.	54	d.	897	f.	965	h.	703	j.	4007
	$\times 69$		$\times 78$		$\times 209$		$\times 376$		$\times 518$

4. Division:

a. $4\overline{)7356}$ **e.** $3\overline{)2851}$ **i.** $36\overline{)1476}$ **m.** $45\overline{)13,954}$

b. $8\overline{)3656}$ **f.** $5\overline{)1702}$ **j.** $86\overline{)6243}$ **n.** $22\overline{)4533}$

c. $7\overline{)1892}$ **g.** $9\overline{)1443}$ **k.** $25\overline{)5017}$ **o.** $76\overline{)7676}$

d. $6\overline{)7818}$ **h.** $7\overline{)5653}$ **l.** $67\overline{)63,650}$ **p.** $58\overline{)33,046}$

Chapter 2
Introduction to Algebra

2-1 Comparing Numbers Using Symbols of Equality and Inequality

STUDY GUIDE

When two numbers are compared, one number is:
(1)   equal to the other, or
(2)   greater than the other, or
(3)   less than the other.

Symbols show the relationship between two quantities. The table below shows three mathematical symbols and the way they are read.

Symbol	Read:
=	Is equal to
<	Is less than
>	Is greater than

A mathematical sentence that uses the symbol "=" is called an equation. A mathematical sentence that uses the symbol "<" or ">" is called an inequality.

FACT FINDING

1. The symbol _____ is used to show that two numerals name the same number.

2. The symbol "<" means _____ .

3. The symbol ">" means _____ .

4. A mathematical sentence that uses the symbol "<" or ">" is called an _____ .

MODEL PROBLEM

Write each of the following mathematical sentences in words:

	Mathematical Sentence	Statement in Words
a.	8 < 10	8 is less than 10.
b.	7 > 3	7 is greater than 3.
c.	6 = 6	6 is equal to 6.

1. Express each statement below using symbols.

a. 4 is less than 16.

b. 5 is greater than zero.

c. 10 is equal to 10.

d. 22 is more than 17.

e. 38 is less than 40.

f. 12 is equal to the quotient of 24 and 2.

g. Twelve is greater than nine.

h. Ten minus nine is greater than zero.

i. Thirty is less than four times sixty.

j. Sixteen increased by three is less than twenty-five.

k. Five times eight is less than fifty minus six.

l. The product of seven and eight is equal to fifty-six.

m. The quotient of ninety-four and two is less than the difference between eighty and fifteen.

n. Eighteen divided by three is equal to eight minus two.

o. The sum of fifty-two and seventeen is less than ninety-two diminished by thirteen.

p. Twenty-three less than two hundred is greater than the product of 32 and 5.

q. One hundred three decreased by forty-seven is equal to 224 divided by 4.

r. Seventy-four diminished by the product of 3 and 10 is less than twice the sum of 12 and 13.

2. Write each of the following statements in words:

a. $11 = 11$

b. $4 > 0$

c. $15 < 24$

d. $39 + 16 = 11 \cdot 5$

e. $14 - 5 = 3 \times 3$

f. $100 \div 25 > 3(1)$

g. $(19)(16) < 500 - 89$

h. $106 + 59 < 924 \div 4$

i. $6 \cdot 0 = 4 - 4$

j. $3c > 18$

k. $25 < x + 8$

l. $7t + 13 > 60$

m. $\dfrac{3m}{4} < 30$

n. $5y - 8 = 2$

3. For each statement below write another inequality having the same meaning as the given one.

a. $14 < 21$ Answer: $21 > 14$

b. $0 < 18$ **c.** $39 > 29$ **d.** $12 < 22$ **e.** $100 > 99$

4. Tell whether each of the following statements is TRUE or FALSE:

a. $14 > 12$

b. $38 < 35$

c. $16 + 7 = 20 + 3$

d. $17 + 6 + 5 < 16 + 3 + 7$

e. $6 + 0 < 6$

f. $131 - 19 > 103 + 9$

g. $4 + 6 + 8 > 6 \cdot 2$

h. $4 + 4 = 4(4)$

i. $\dfrac{180}{30} = 5 + 1$

j. $5 < 7 < 8$

k. $10 > 6 > 7$

l. $4(8) - 3 < 50 \div 2 + 1$

m. $14 \times 2^2 = 7 \times 2^3$

n. $97 + 3 < 10^2$

o. $(10 - 3)(10 + 3) = 10^2 - 3^2$

5. Write the symbol of equality or inequality that makes each of the following statements true:

a. $19 \underline{} 7$

b. $5 \underline{} 13$

c. $1 - 0 \underline{} 1 \times 0$

d. $6 + 1 \underline{} 9$

e. $13 \underline{} 8 + 5$

f. $6 + 6 \underline{} 30$

g. $61 \underline{} 48 + 7$

h. $13 + 9 + 20 \underline{} 16 + 7$

i. $23 + 4 \underline{} 9 + 3$

j. $17 + 2 \underline{} 24 - 6$

k. $42 \div 7 \underline{} 48 \div 6$

l. $8(5) \underline{} 63 - 34$

m. $24 + 58 \underline{} 82 - 10$

n. $33 + 11 \underline{} 55 - 9$

o. $6 \div 2 \underline{} 9 \div 3$

p. $2^4 \cdot 5 \underline{} 2^3 \cdot 10$

q. $3^2(10) \underline{} 3(10^2)$

r. $(12 - 7) - 5 \underline{} 12 - (7 - 5)$

s. $10^2 - 8^2 \underline{} 36$

t. $(9 + 3)(4 - 2) \underline{} 24$

u. $(27 \div 9) \div 3 \underline{} 27 \div (9 \div 3)$

v. $(8 - 2)(8 + 2) \underline{} 60$

w. $(5 + 4)^2 \underline{} 5^2 + 4^2$

x. $(9 - 5)^2 \underline{} 9^2 - 5^2$

y. $5^2 - 3^2 \underline{} (5 - 3)(5 + 3)$

z. $\dfrac{5(12 + 8)}{2} \underline{} \dfrac{7(12 - 9)}{3}$

6. Write the symbol of equality or inequality that makes each of the following statements true. (Keep in mind the order of operations.)

a. $(24 \div 8) \cdot 8 \underline{} 3^3$

b. $2(3) + 7 \underline{} 10 + 6 \div 2$

c. $3 - (8 - 5) \underline{} 8 \div 8 - 1$

d. $(16 - 2) + 7 \underline{} 8 \times 6 + 12 \div 6$

e. $(5 \cdot 4 - 2) \div 6 \underline{} 4^2$

f. $14 - (2 \cdot 1) + 4 \underline{} 17 - (2 \cdot 4) + 3$

7. If $y = 3$, fill in the blank using the symbol ($<$, $>$, or $=$) that will make each statement true.

  a. $y + 7 \_\_ 20 + 2$

   Solution:
   (1) Substitute $y = 3$: $3 + 7 \_\_ 20 + 2$
   (2) Evaluate: $10 \_\_ 22$
   (3) Compare: $10 < 22$

  b. $3y \_\_ 4$ **k.** $3y - 7 \_\_ 5y + 11$

  c. $4y \_\_ 12$ **l.** $3y - 2 \_\_ 10 + y$

  d. $\dfrac{2y}{3} \_\_ 3$ **m.** $3y + 5 \_\_ 76 - 10$

  e. $3y - 2 \_\_ 5$ **n.** $11y - y \_\_ 25 - 2y$

  f. $8y - 6 \_\_ 7$ **o.** $3y - 2y + 8 \_\_ 5y - 12$

  g. $2y + 4 \_\_ 18$ **p.** $3(y + 1) \_\_ 12$

  h. $3y + y \_\_ 4y$ **q.** $9(y - 3) \_\_ 36$

  i. $2y + 9 \_\_ y + 5$ **r.** $3y \_\_ 4(2y + 15)$

  j. $4y - 9 \_\_ 2y - 1$ **s.** $(y + 3)y \_\_ 49$

8. If x is a whole number, find two values of x that make each of the following statements true:

  a. $x + 8 < 11$ Answer: $x = 1$ since $1 + 8 < 11$, or $9 < 11$.
                                       $x = 2$ since $2 + 8 < 11$, or $10 < 11$.

  b. $x - 4 < 8$ **e.** $23 + x < 31$ **h.** $2x + 5 < 12$

  c. $7x < 26$ **f.** $20 \div 5 > x$ **i.** $3x + 4 < 16$

  d. $2x < 17$ **g.** $17 - x > 12$ **j.** $5 + 2x < 19$

**EXTRA
FOR
EXPERTS**

1. Five more than twice a number is less than 13. What whole numbers make this true?

2. Four less than 3 times a number is greater than 11. What whole numbers make this true?

2-2.1 *The Number Line and Whole Numbers*

STUDY GUIDE   A line may be thought of as an endless set of points. A *number line* is a straight line that has its points labeled with numerals. The set of numbers 0, 1, 2, 3, and so on is called the set of *whole numbers*. All of the numbers shown on the number line below are called whole numbers.

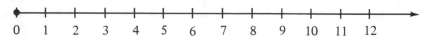

Each whole number may be associated with a point on the number line. No point has two numbers assigned to it, and each number is assigned to only one point. In other words, the points and the numbers are in a one-to-one correspondence. Of two numbers on a number line, the one on the right is the greater number.

> *Example:*   7 is to the right of 5.
> Therefore 7 is the greater number.
> This may be written as $7 > 5$ or $5 < 7$.

FACT FINDING

1.   A line upon which positions represent numbers is called a _____ _____ .

2.   The set of whole numbers begins with _____ .

3.   On a number line each point is associated with one number, and each number is associated with one point. This is called a _____ correspondence.

4.   On the number line, the greater of two numbers is located to the _____ of the smaller number.

5.   On the number line, the _____ of two numbers is located on the left.

6.   If *a* is to the right of *b*, then _____ or _____ .

EXERCISES

1.   What number is:

a.   5 units to the right of 2?

b.   4 units to the left of 9?

c.   3 units to the right of 0?

d.   3 units to the left of 10?

e.   6 units to the right of 2?

f.   3 units to the left of 8?

g.   0 unit to the right of 6?

h.   2 units to the left of 7?

i.   7 units to the right of 1?

j.   0 unit to the left of 5?

2. a. Locate point 5 on the number line. How many units is it from point 9? In which direction?

b. Locate point 2 on the number line. How many units is it from point 7? In which direction?

c. Locate point 3 on the number line. How many units is point 3 from point 9? In which direction?

d. Locate point 8 on the number line. How many units is point 8 from point 3? In which direction?

e. Locate point 12 on the number line. How many units is point 12 from point 5? In which direction?

3. For each pair of numbers, name the number that will appear farther to the right on a number line.

a. 5, 7	**d.** 8, 5	**g.** 493, 492
b. 9, 4	**e.** 0, 6	**h.** 1, 4
c. 3, 1	**f.** 922, 453	**i.** 5, 0

4. Use an inequality symbol to show the relation between the first and the second number listed in each of the following pairs:

a. 7, 9	**e.** 8, 2	**i.** 10, 2
b. 8, 4	**f.** 6, 11	**j.** 12, 15
c. 4, 1	**g.** 1, 3	**k.** 15, 12
d. 9, 5	**h.** 9, 6	**l.** 0, 1

5. Using the number line pictured below, write the symbol ($<$, $>$, or $=$) that should be used in the space to make each resulting statement true.

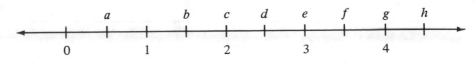

a. a __ d	**d.** 0 __ e	**g.** b __ 1
b. 2 __ c	**e.** f __ a	**h.** g __ 4
c. 1 __ d	**f.** h __ 0	

2-2.2 *The Number Line and Rational Numbers*

STUDY GUIDE In a fraction, the number written below the fraction bar is called the denominator (it is the number of equal parts into which each unit is divided). The number written above the fraction bar is called the numerator (it is the number of equal parts taken).

Therefore a fraction may be defined as:

$$\frac{\text{Number of parts taken}}{\text{Number of parts in the whole}}.$$

Fractions that name numbers between 0 and 1 are called *proper fractions*. A proper fraction has a numerator that is less than the denominator.

Example: $\frac{3}{4}$ is a proper fraction since $\frac{3}{4}$ is between 0 and 1.

A *mixed number* is a number consisting of a whole number plus a proper fraction.

Example: $2\frac{1}{4}$ is a mixed number since $2\frac{1}{4} = 2 + \frac{1}{4}$.

The set of *rational numbers* includes whole numbers, proper fractions, and mixed numbers. There is a point on the number line corresponding to each whole number, proper fraction, and mixed number. Fractions may be compared by using the number line. The larger fraction is the one whose position on the number line is farther to the right.

FACT FINDING

1. In the fraction $\frac{3}{4}$, _____ is the numerator and 4 is the _____ .

2. The denominator of a fraction names the number of _____ _____ in the _____ .

3. A fraction may be defined as _____ _____ _____ _____ divided by _____ _____ _____ _____ _____ _____ .

4. A numeral such as $3\frac{3}{4}$ is called a _____ number.

5. A fraction whose numerator is less than its denominator is called a _____ fraction.

6. On the number line, the greater of two fractions is located to the _____ of the smaller fraction.

1. For each pair of numbers, name the number that will appear farther to the right on a number line.

a. $1, 1\frac{1}{4}$ d. $5\frac{3}{4}, 5\frac{1}{4}$ g. $\frac{4}{4}, 1\frac{3}{4}$

b. $5, 7\frac{1}{4}$ e. $\frac{1}{4}, \frac{3}{4}$ h. $\frac{1}{2}, \frac{8}{8}$

c. $7\frac{1}{2}, 9$ f. $4\frac{1}{4}, 7\frac{1}{4}$ i. $2, \frac{4}{4}$

2. Which of the following name a whole number?

a. $\frac{3}{5}$ b. 4 c. $5\frac{1}{2}$ d. 9 e. $\frac{7}{8}$ f. $2\frac{3}{4}$

3. Which of the following name a proper fraction?

a. $\frac{2}{5}$ b. $\frac{7}{4}$ c. $\frac{8}{8}$ d. $\frac{1}{3}$ e. $\frac{7}{8}$ f. $\frac{6}{5}$

4. Which of the following name a mixed number?

a. 4 b. $2\frac{2}{3}$ c. $\frac{5}{7}$ d. $4\frac{1}{5}$ e. 6 f. $1\frac{7}{16}$

**EXTRA
FOR
EXPERTS**

1. Locate the point $1\frac{1}{4}$ on the number line. Count $\frac{3}{4}$ unit to the right; count $4\frac{3}{4}$ units to the right; count $2\frac{2}{4}$ units to the left; count $1\frac{3}{4}$ units to the left; count $2\frac{1}{4}$ units to the right. At what point on the number line are you?

2. How many $\frac{1}{4}$ inches are there in 3 inches?

3. How many $\frac{1}{4}$ inches are there in $5\frac{1}{4}$ inches?

4. Find the distance between $\frac{1}{4}$ and $3\frac{3}{4}$ inches.

5. In each of the following, replace the question mark by a number so that the resulting statement will be true:

a. $2 = \frac{?}{4}$ b. $\frac{1}{2} = \frac{?}{4}$ c. $3\frac{1}{2} = \frac{?}{4}$

2-3.1 *Ruler Fractions*

A commonly used tool for measuring length is the *ruler*. Some rulers use the unit of linear measure called the inch. Inch is abbreviated as in. The symbol ″ means inch.

Look at an enlargement of 1 in. on a ruler. The divisions of the inch are marked. The line marked $\frac{1}{2}$ divides the inch into halves. The $\frac{1}{4}$-inch mark is found halfway between the left edge of the ruler and the $\frac{1}{2}$-inch mark. The $\frac{1}{8}$-inch mark is found halfway between the left edge and the $\frac{1}{4}$-inch mark.

$\frac{1}{8}$ $\frac{1}{4}$ $\frac{3}{8}$ $\frac{1}{2}$ $\frac{5}{8}$ $\frac{3}{4}$ $\frac{7}{8}$ $\frac{2}{2}$ $\frac{9}{8}$

$\frac{2}{8}$ $\frac{2}{4}$ $\frac{6}{8}$ $\frac{4}{4}$

$\frac{4}{8}$ $\frac{8}{8}$

1

In a similar manner, we can further divide the 1 inch into 16 equal parts by considering intervals equal in length to the interval halfway between the left edge of the ruler and the $\frac{1}{8}$-inch mark.

The meaning of each mark on a ruler may be determined by counting the number of equal parts into which the whole inch has been divided.

FACT FINDING

1. The inch is one unit of _____ measure.

2. The abbreviation for inch is _____ .

3. If each 1-inch length is divided into two equal lengths, each of these equal parts is _____ of an inch.

4. If each 1-inch length is divided into four equal lengths, each of these equal parts is _____ of an inch.

5. If each 1-inch length is divided into _____ equal lengths, each of these equal parts is $\frac{1}{8}$ of an inch.

6. If each 1-inch length is divided into _____ equal lengths, each of these equal parts is $\frac{1}{16}$ of an inch.

7. By counting the number of equal parts into which the whole inch has been divided, we can determine the _____ of each mark.

MODEL PROBLEM

Using a ruler, change the fraction $\frac{1}{2}$ into an equivalent fraction with 16 as its denominator.

Solution: (1) Using the ruler in the Study Guide, measure the space from the left edge to the first unmarked division. It is $\frac{1}{16}$ inch.

(2) Count the number of these spaces between the left edge of the ruler and $\frac{1}{2}$. There are 8 of these spaces.

(3) Answer: $\frac{1}{2} = \frac{8}{16}$.

EXERCISES

1. Change each of the following fractions to an equivalent fraction, using a ruler:

a. $\frac{1}{2} = \frac{}{8}$

b. $\frac{1}{4} = \frac{}{16}$

c. $\frac{3}{4} = \frac{}{16}$

d. $\frac{3}{4} = \frac{}{8}$

e. $\frac{7}{8} = \frac{}{16}$

f. $\frac{5}{8} = \frac{}{16}$

g. $\frac{3}{8} = \frac{}{16}$

h. $\frac{1}{8} = \frac{}{16}$

i. $\frac{2}{2} = \frac{}{4}$

j. $\frac{4}{4} = \frac{}{16}$

k. $\frac{1}{4} = \frac{}{8}$

l. $\frac{8}{8} = \frac{}{16}$

2. Using the ruler drawn:

a. Determine the meaning of each division.

b. Change each fraction given below to an equivalent fraction.

(1) $\frac{1}{2} = \frac{}{12}$

(2) $\frac{2}{4} = \frac{}{12}$

(3) $\frac{1}{4} = \frac{}{12}$

(4) $\frac{3}{4} = \frac{}{12}$

(5) $\frac{4}{4} = \frac{}{12}$

(6) $\frac{6}{8} = \frac{}{12}$

3. Using the ruler drawn:

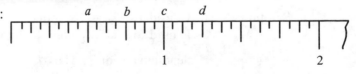

a. Give three names for point a.

b. Give three names for point b.

c. Give three names for point c.

d. Give three names for point d.

2-3.2 *Equivalent Fractions*

Using the ruler in the preceding Study Guide, we can see that $\frac{1}{2}$ inch, $\frac{2}{4}$ inch, and $\frac{4}{8}$ inch all name the same number. *Equivalent fractions* are fractions that name the same number. We can change a fraction into an equivalent fraction without the use of a ruler by multiplying the numerator and the denominator of the fraction by the same nonzero number.

FACT FINDING

1. Fractions that name the same number are called _____ fractions.

2. To obtain the fraction $\frac{6}{8}$ from the equivalent fraction $\frac{3}{4}$, you _____ the numerator and the denominator of $\frac{3}{4}$ by _____.

3. The fractions $\frac{2}{2}, \frac{3}{3}, \frac{4}{4}, \frac{5}{5}$, and $\frac{6}{6}$ are each equivalent to _____.

MODEL PROBLEMS

1. Change the fraction $\frac{1}{2}$ into an equivalent fraction with a denominator of 16.

   Solution: (1) Since our rule says we must *multiply* both the numerator and the denominator by the same nonzero number, consider 2(?) = 16. Thus ? = 8.

   (2) Answer: $\frac{1}{2} = \frac{1 \cdot 8}{2 \cdot 8} = \frac{8}{16}$.

2. Change the fraction $\frac{5}{4}$ into an equivalent fraction with a denominator of 16.

   Solution: (1) Consider 4(?) = 16 to find the nonzero number that is used to multiply both the numerator and the denominator of $\frac{5}{4}$. Thus ? = 4.

   (2) Answer: $\frac{5}{4} = \frac{5 \cdot 4}{4 \cdot 4} = \frac{20}{16}$.

1. Change each of the following fractions to an equivalent fraction by using the multiplication principle:

a. $\dfrac{3}{4} = \dfrac{}{16}$

e. $\dfrac{7}{8} = \dfrac{}{32}$

i. $\dfrac{1}{4} = \dfrac{}{32}$

b. $\dfrac{3}{4} = \dfrac{}{8}$

f. $\dfrac{4}{4} = \dfrac{}{8}$

j. $\dfrac{1}{8} = \dfrac{}{32}$

c. $\dfrac{7}{8} = \dfrac{}{16}$

g. $\dfrac{3}{4} = \dfrac{}{12}$

k. $\dfrac{1}{16} = \dfrac{}{32}$

d. $\dfrac{1}{2} = \dfrac{}{100}$

h. $\dfrac{13}{16} = \dfrac{}{80}$

l. $\dfrac{3}{4} = \dfrac{}{32}$

2. For each question mark substitute the numeral that will make the fractions equivalent.

a. $\dfrac{?}{4} = \dfrac{12}{16}$

c. $\dfrac{?}{50} = \dfrac{74}{100}$

e. $\dfrac{9}{?} = \dfrac{18}{16}$

b. $\dfrac{17}{30} = \dfrac{?}{180}$

d. $\dfrac{?}{90} = \dfrac{108}{360}$

f. $\dfrac{7}{?} = \dfrac{56}{64}$

2-3.3 *Comparing Ruler Fractions*

STUDY GUIDE To compare two ruler fractions, recall that the larger fraction is farther to the right on the ruler.

Example: $\dfrac{1}{4} < \dfrac{3}{4}$ or $\dfrac{3}{4} > \dfrac{1}{4}$ since $\dfrac{3}{4}$ is farther to the right than $\dfrac{1}{4}$.

Ruler fractions can also be compared without the use of a ruler by using the following fact: If two fractions have the same denominator, the fraction with the greater numerator names the greater fraction.

Example: $\dfrac{1}{4}$ and $\dfrac{3}{4}$ have the same denominator, and since 3 is greater than 1, $\dfrac{3}{4}$ is greater than $\dfrac{1}{4}$ $\left(\text{or } \dfrac{3}{4} > \dfrac{1}{4}\right)$.

If two fractions have unlike denominators, first change the fractions into equivalent fractions so that they have the same denominator, and then compare their numerators.

FACT FINDING

1. On the ruler, the smaller of two fractions is located to the

_____ of the larger fraction.

2.   When two fractions having the same denominator are compared, the fraction with the larger _____ is the larger fraction.

3.   If two fractions having different denominators are to be compared, the fractions must first be changed into _____ fractions having the same _____ .

MODEL PROBLEM

Compare $\frac{1}{4}$ and $\frac{1}{8}$.

Solution: (1)   Change $\frac{1}{4}$ into an equivalent fraction having 8 as denominator. Consider $4(?) = 8$. Thus $? = 2$.

(2)   Multiply the numerator and denominator of $\frac{1}{4}$ by 2 to obtain an equivalent fraction with a denominator of 8. Thus

$$\frac{1}{4} = \frac{1 \cdot 2}{4 \cdot 2} = \frac{2}{8}.$$

(3)   Since $\frac{2}{8}$ and $\frac{1}{8}$ have the same denominator, compare the numerators. Since 2 is greater than 1, $\frac{2}{8} > \frac{1}{8}$.

(4)   Answer: $\frac{1}{4} > \frac{1}{8}$.

EXERCISES

1.   Place the symbol "<" or ">" between the two numerals so that the resulting statement will be true.

a. $\frac{5}{9} - \frac{2}{9}$ c. $\frac{25}{36} - \frac{30}{36}$ e. $\frac{33}{150} - \frac{24}{150}$

b. $\frac{8}{12} - \frac{10}{12}$ d. $\frac{75}{100} - \frac{50}{100}$ f. $\frac{75}{90} - \frac{66}{90}$

2.   Arrange the fractions so that the rational numbers that they represent will appear in order of value from the least value to the greatest value.

a. $\frac{5}{8}, \frac{1}{8}, \frac{9}{8}, \frac{7}{8}$ c. $\frac{1}{10}, \frac{3}{10}, \frac{7}{10}, \frac{5}{10}$ e. $\frac{5}{12}, \frac{11}{12}, \frac{7}{12}, \frac{1}{12}$

b. $\frac{5}{3}, \frac{8}{3}, \frac{1}{3}, \frac{4}{3}$ d. $\frac{4}{5}, \frac{2}{5}, \frac{3}{5}, \frac{6}{5}$ f. $\frac{7}{16}, \frac{3}{16}, \frac{10}{16}, \frac{15}{16}$

3. Rename each pair of fractions so that they have like denominators.

a. $\frac{1}{2}$ and $\frac{3}{8}$ **d.** $\frac{3}{8}$ and $\frac{1}{4}$ **g.** $\frac{3}{10}$ and $\frac{13}{100}$

b. $\frac{7}{4}$ and $\frac{3}{2}$ **e.** $\frac{3}{4}$ and $\frac{3}{8}$ **h.** $\frac{3}{8}$ and $\frac{11}{32}$

c. $\frac{3}{16}$ and $\frac{1}{4}$ **f.** $\frac{7}{8}$ and $\frac{13}{16}$ **i.** $\frac{6}{5}$ and $\frac{16}{15}$

4. Place the symbol ">," "<," or "=" between the two numerals so that the resulting statement will be true. Use your ruler to check your answers.

a. $\frac{3}{4} \underline{\quad} \frac{1}{2}$ **d.** $\frac{5}{8} \underline{\quad} \frac{7}{16}$ **g.** $\frac{5}{8} \underline{\quad} \frac{11}{16}$

b. $\frac{3}{4} \underline{\quad} \frac{11}{16}$ **e.** $\frac{1}{2} \underline{\quad} \frac{8}{16}$ **h.** $\frac{3}{16} \underline{\quad} \frac{1}{8}$

c. $\frac{3}{4} \underline{\quad} \frac{12}{16}$ **f.** $\frac{7}{8} \underline{\quad} \frac{1}{2}$ **i.** $\frac{4}{8} \underline{\quad} \frac{1}{2}$

2-3.4 *Changing Mixed Numbers to Improper Fractions*

STUDY GUIDE The ruler shown below can be used to find the number of times $\frac{1}{2}$ inch is contained in $2\frac{1}{2}$ inches.

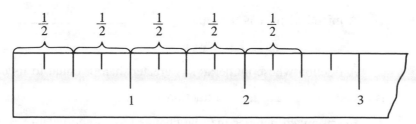

There are five $\frac{1}{2}$-inch units contained in the number $2\frac{1}{2}$. Therefore $2\frac{1}{2}$ is equivalent to $\frac{5}{2}$ or $2\frac{1}{2}'' = \frac{5''}{2}$.

A mixed number was defined as the sum of a whole number and a proper fraction. Therefore:

$$2\frac{1}{2} = 2 + \frac{1}{2} = 1 + 1 + \frac{1}{2}$$

$$= \frac{2}{2} + \frac{2}{2} + \frac{1}{2}$$

$$= \frac{5}{2}.$$

This illustrates one method of changing a mixed number $\left(\text{such as } 2\frac{1}{2}\right)$ to an improper fraction $\left(\frac{5}{2}\right)$. Fractions that name numbers equal to 1 or greater than 1 are called *improper fractions*. Therefore an improper fraction $\left(\text{such as } \frac{5}{2}\right)$ is a fraction in which the numerator is equal to or greater than the denominator.

Another method that can be used to change a mixed number to an improper fraction (without using a ruler) is outlined below.

1. Using the definition of a mixed number, write

$$2\frac{1}{2} = 2 + \frac{1}{2}.$$

2. Rewriting the whole number 2 as $\frac{2}{1}$, change $\frac{2}{1}$ to an equivalent fraction with a denominator of 2 by multiplying both the numerator and the denominator of $\frac{2}{1}$ by 2:

$$\frac{2}{1} = \frac{2 \cdot 2}{1 \cdot 2} = \frac{4}{2}$$

3. Thus:

$$2\frac{1}{2} = 2 + \frac{1}{2}$$

$$= \frac{4}{2} + \frac{1}{2}$$

$$= \frac{5}{2}.$$

FACT FINDING

1. A mixed number is the sum of _____ _____

_____ and _____ _____ _____ .

2. A fraction whose numerator is greater than or equal to its denominator is called an _____ fraction.

3. Every mixed number may be changed to an equivalent

_____ fraction.

4. A whole number may be written as a fraction with a denominator of

_____ .

MODEL PROBLEMS

1. Using the ruler, find the number of $\frac{1}{4}$ inches contained in $3\frac{1}{4}$ inches.

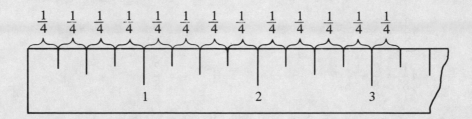

Solution: The unit $\frac{1}{4}''$ is contained 13 times in $3\frac{1}{4}''$.

Answer: $3\frac{1}{4} = \frac{13}{4}$.

2. Without using the ruler, change the mixed number $3\frac{1}{4}$ to an improper fraction.

Solution: $3\frac{1}{4} = 3 + \frac{1}{4}$

$= \frac{12}{4} + \frac{1}{4}$ $\left(\text{since } 3 = \frac{3}{1} = \frac{3 \cdot 4}{1 \cdot 4} = \frac{12}{4}\right)$

$= \frac{13}{4}$.

Answer: $3\frac{1}{4} = \frac{13}{4}$.

EXERCISES

1. Use a ruler to determine how many:

a. $\frac{1}{4}$ inches are in $1\frac{1}{4}$ inches

b. $\frac{1}{8}$ inches are in $\frac{3}{4}$ of an inch

c. $\frac{1}{2}$ inches are in $3\frac{1}{2}$ inches

d. $\frac{1}{16}$ inches are in $1\frac{1}{4}$ inches

e. $\frac{1}{8}$ inches are in 2 inches

f. $\frac{1}{4}$ inches are in $2\frac{1}{2}$ inches

2. Find the number that can replace the question mark so that the statement will be true.

a. $4\frac{1}{2} = 4 + \frac{1}{2} = \frac{?}{2} + \frac{1}{2}$

g. $7\frac{3}{8} = \frac{56}{?} + \frac{3}{8}$

b. $2\frac{3}{4} = 2 + \frac{3}{4} = \frac{?}{4} + \frac{3}{4}$

h. $12\frac{1}{2} = \frac{24}{?} + \frac{1}{2}$

c. $5\frac{3}{8} = \frac{?}{8} + \frac{3}{8}$

i. $1 = \frac{?}{20}$

d. $4\frac{1}{8} = \frac{?}{8} + \frac{1}{8}$

j. $4 = \frac{?}{8}$

e. $3\frac{5}{16} = \frac{?}{16} + \frac{5}{16}$

k. $2 = \frac{?}{12}$

f. $2\frac{5}{16} = \frac{32}{16} + \frac{?}{16}$

l. $7 = \frac{?}{16}$

3. Express each of the following mixed numbers as an improper fraction:

a. $3\frac{3}{4}$ d. $15\frac{2}{3}$ g. $17\frac{5}{7}$ j. $14\frac{5}{6}$ m. $3\frac{4}{25}$

b. $8\frac{7}{8}$ e. $3\frac{5}{7}$ h. $11\frac{3}{5}$ k. $16\frac{8}{9}$ n. $3\frac{2}{9}$

c. $13\frac{1}{4}$ f. $19\frac{1}{2}$ i. $6\frac{3}{10}$ l. $4\frac{7}{12}$ o. $7\frac{1}{16}$

2-3.5 *Changing Improper Fractions to Mixed Numbers*

STUDY GUIDE The picture of the ruler shows that $\frac{4}{4} = 1''$.

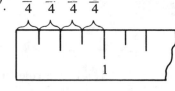

The point $\frac{11''}{4}$ from the left edge of the ruler is also the point named $2\frac{3''}{4}$.

This means that $\frac{11''}{4} = 2\frac{3''}{4}$.

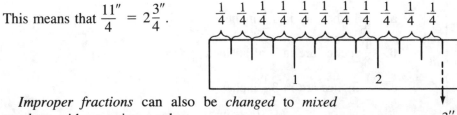

Improper fractions can also be *changed* to *mixed numbers* without using a ruler.

Example: $\frac{11}{4} = \frac{4}{4} + \frac{4}{4} + \frac{3}{4}$ $= 1 + 1 + \frac{3}{4}$ $= 2\frac{3}{4}$.

The technique most commonly used to change an improper fraction to a mixed number (without using a ruler) is to divide the numerator of the fraction by the denominator.

MODEL PROBLEM

Change $\frac{11}{4}$ to a mixed number.

Solution: (1) Write the improper fraction as a division example:

$$\frac{11}{4} = 4\overline{)11}.$$

(2) Divide to get the whole-number part of the mixed number:

$$\overset{2\,R\,3}{4\overline{)11}}.$$

(3) Write the remainder as the numerator of the fraction part of the mixed number, and the divisor as the denominator of the fraction part of the mixed number:

$$\overset{2\frac{3}{4}}{4\overline{)11}}.$$

(4) Answer: $\frac{11}{4} = 2\frac{3}{4}$.

EXERCISES

1. Which of the following name an improper fraction?

   a. $\frac{1}{2}$ c. $\frac{18}{6}$ e. $\frac{8}{5}$ g. $\frac{22}{7}$ i. $\frac{8}{8}$ k. $\frac{21}{12}$

   b. $\frac{10}{10}$ d. $\frac{5}{8}$ f. $\frac{2}{3}$ h. $\frac{9}{4}$ j. $\frac{15}{20}$ l. $\frac{32}{32}$

2. Use a ruler to determine the equivalent whole number or mixed number.

   a. $\frac{2''}{2}$ d. $\frac{16''}{8}$ g. $\frac{5''}{2}$ j. $\frac{37''}{16}$ m. $\frac{14''}{8}$

   b. $\frac{4''}{2}$ e. $\frac{32''}{16}$ h. $\frac{23''}{8}$ k. $\frac{6''}{4}$ n. $\frac{10''}{8}$

   c. $\frac{8''}{2}$ f. $\frac{5''}{4}$ i. $\frac{25''}{16}$ l. $\frac{20''}{8}$ o. $\frac{42''}{16}$

3. Express each of the following as a whole number or a mixed number:

a. $\dfrac{5}{5}$ d. $\dfrac{56}{8}$ g. $\dfrac{8}{7}$ j. $\dfrac{15}{8}$ m. $\dfrac{35}{8}$

b. $\dfrac{10}{2}$ e. $\dfrac{48}{8}$ h. $\dfrac{3}{2}$ k. $\dfrac{20}{7}$ n. $\dfrac{13}{10}$

c. $\dfrac{32}{4}$ f. $\dfrac{20}{4}$ i. $\dfrac{75}{4}$ l. $\dfrac{49}{16}$ o. $\dfrac{20}{3}$

4. Write the symbol "<" or ">" so that each statement will be true.

a. $1\dfrac{4}{7} - \dfrac{10}{7}$ Solution: $\dfrac{10}{7} = 1\dfrac{3}{7}$. Therefore $1\dfrac{4}{7} > 1\dfrac{3}{7}$.

b. $4\dfrac{5}{7} - \dfrac{29}{7}$ d. $\dfrac{34}{5} - 6\dfrac{3}{5}$ f. $5\dfrac{3}{10} - \dfrac{51}{10}$

c. $\dfrac{46}{9} - 5\dfrac{2}{9}$ e. $\dfrac{20}{3} - 6\dfrac{1}{3}$ g. $\dfrac{10}{3} - 6\dfrac{2}{3}$

5. Express your answers to the following problems as mixed numbers:

a. A board 11 feet long is divided into 4 equal lengths. How long is each part?

b. A typist types 5 pages in an hour. How many hours will she take to type 123 pages?

c. A truckload of dresses contains 364 dresses. How many dozen dresses are there?

d. A class of 34 students made a total of 80 errors on a test. What was the average number of errors made per student?

e. Ten pies are to be divided equally among 3 people. How much will each one get?

EXTRA FOR EXPERTS

1. Find the $1\dfrac{3}{4}''$ mark on the ruler. Add $2\dfrac{1}{2}''$. Subtract $\dfrac{5}{8}''$. What is the result?

2. Find the $5\dfrac{7}{8}''$ mark on the ruler. Subtract $4\dfrac{1}{8}''$. Add $1\dfrac{1}{2}''$. What is the result?

2-4.1 *Reading a Ruler*

STUDY GUIDE The meaning of each mark on a ruler can be determined by counting the number of equal parts into which a whole inch has been divided. The left edge of a ruler is the zero point on the ruler.

Examples:

There are 8 spaces to the inch. Therefore each space is $\frac{1}{8}$ of an inch.

The distance from the left edge of the ruler to *A* is 3 spaces or $\frac{3}{8}$ of an inch.

The distance from the left edge to *B* is 6 spaces or $\frac{6}{8}$ of an inch. It is also $\frac{3}{4}$ of an inch. $\frac{3}{4}$ is called the simplest name of point *B*.

The distance from the left edge to *C* is 13 spaces or $\frac{13}{8}$ of an inch or $1\frac{5}{8}$ inches.

FACT FINDING

1. If each 1-inch length is divided into 16 equal lengths, each of these lengths is _____ of an inch.

2. To find a distance on a ruler, start counting from the _____ edge of the ruler.

3. The _____ edge of a ruler is the zero point on the ruler.

EXERCISES

1. Give the simplest name of every point pictured on the ruler below:

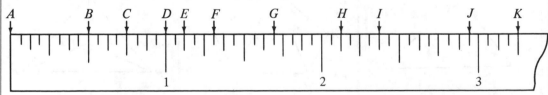

2. Using the ruler in Exercise 1, find the distance on the ruler from:

a. *B* to *D*	**d.** *D* to *G*	**g.** *F* to *G*	**j.** *D* to *J*
b. *C* to *D*	**e.** *G* to *H*	**h.** *F* to *H*	**k.** *G* to *K*
c. *B* to *E*	**f.** *D* to *H*	**i.** *B* to *I*	**l.** *J* to *K*

3. Using the ruler in Exercise 1, name any two letters that are:

a. 1 inch apart   **b.** $\frac{1}{4}$ inch apart   **c.** $\frac{1}{2}$ inch apart

2-4.2 *Line Segments: Naming, Measuring*

STUDY GUIDE A capital letter is used to label and name a *point:*

. *R* is "point *R*."

A *line* is represented as: 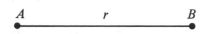 . The arrowheads are used to show that a line is endless in both directions.

A line has an infinite set of points but no endpoints.

A *line segment* is the part of a line between two of its points. It is represented by the capital letters of these points or by a small letter.

A ●————— *r* —————● B

Thus $\overline{AB}$ or *r* represents the line segment between *A* and *B*.

FACT FINDING

1. A line has _____ endpoints.

2. A line segment has _____ endpoints.

3. A point is named by _____ capital letter.

4. A line segment is named by _____ capital letters.

5. A _____ may be extended indefinitely in either direction.

EXERCISES

1. Name all the line segments in each figure below.

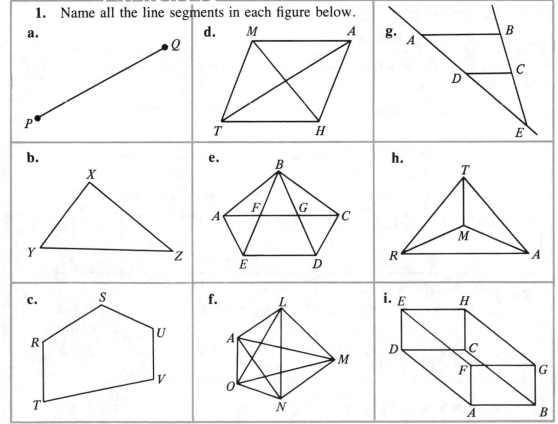

a. **b.** **c.** **d.** **e.** **f.** **g.** **h.** **i.**

2. Use your ruler to find the length to the nearest eighth of an inch of each of the following line segments. Express each length by its simplest name.

a.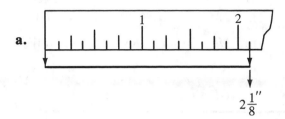

$$2\frac{1}{8}''$$

Solution: (1) Place the left edge of your ruler at the beginning of the line as shown.

(2) Read the point on the ruler corresponding to the end of the line.

Answer: The length of this line segment is $2\frac{1}{8}''$.

b. _____

c. _____

d. _____

e. _____

f. _____

g. _____

h. _____

i. _____

3. Find the measure of each side of the figures shown, and list the names of these sides for each figure in order from smallest to largest.

a.

B

A   C

c.

K

H   L

e.

O

N   P

M   Q

b.

D   E

R   T

d.

Y

X   Z

f.

T   U

S   V

4. Use your ruler to find the lengths to the nearest eighth of an inch of the following line segments shown in the diagram below:

a.	*AB*	**f.**	*FG*	**k.**	*KL*	**p.**	*PQ*	**t.**	*TU*
b.	*BC*	**g.**	*GH*	**l.**	*LM*	**q.**	*QR*	**u.**	*UV*
c.	*CD*	**h.**	*HI*	**m.**	*MN*	**r.**	*RS*	**v.**	*VW*
d.	*DE*	**i.**	*IJ*	**n.**	*NO*	**s.**	*ST*	**w.**	*WX*
e.	*EF*	**j.**	*JK*	**o.**	*OP*				

1. If each $\frac{1}{2}''$ on a ruler represents 30 miles, how many miles are represented by 3"?

2. If $\frac{1}{8}'' = 1$ foot, how long a line segment would be needed to represent 32 feet?

3. **a.** Express $\overline{RT}$ as the sum of three segments.

    b. Express $\overline{RT}$ as the sum of two segments in two different ways.

4. Express $\overline{RS}$ as the difference between two segments.

2-5 *Describing Changes by Directed Numbers*

STUDY GUIDE Changes can be described as *positive* or *negative*. Some examples of changes are:

Positive Change	Negative Change
Win	Lose
Earn	Spend
Up	Down
Gain	Lose
Above	Below
Deposit	Withdraw

When numbers are used to represent positive changes, they are written with a plus (+) sign before the number.

When numbers are used to represent negative changes, they are written with a minus (−) sign before the number.

The number 0 is used to represent no change.

The positive and negative signs are used to indicate the direction of the change. Positive and negative numbers are called *directed numbers*. Signed numbers are used to indicate:

1. Quantity or amount.
2. Direction of change.

The + sign shows one direction, and the − sign shows the opposite direction. Zero (0) is neither a positive nor a negative number. Directed numbers and 0 are called *signed numbers*.

FACT FINDING

1. A gain can be called a _____ change.

2. Negative changes are written with a _____ sign.

3. No change is represented by _____ .

4. The set of positive and negative numbers is called the set of

_____ numbers.

5. A number with both quantity and direction is called _____ .

6. If the + sign represents one direction, the _____ sign represents the opposite direction.

7. _____ is neither a positive nor a negative number.

1. Express each change as a positive number, a negative number, or 0.

a. 5° below zero	**g.** A loss of $27	**m.** Neither loss nor gain
b. A gain of 15 yards	**h.** 16° above zero	**n.** 8 points below average
c. Earnings of $13	**i.** At sea level	**o.** A saving of $14
d. No change in weight	**j.** A debt of $50	**p.** A $20 deposit in the bank
e. 87 years ago	**k.** 3 steps forward	**q.** A $3 drop in price
f. A profit of $31	**l.** 8 steps backward	**r.** A withdrawal of $10

2. Describe the opposite of each of the following situations, and represent the opposite situation as a signed number:

a. A bank deposit of $56	**h.** 20° above zero
b. A 10-yard loss in football	**i.** 70 feet to the left
c. An increase of $5 in weekly earnings	**j.** An increase of 15%
d. 17 feet below sea level	**k.** 4 miles east
e. 3 miles north	**l.** Saved $14
f. A.D. 25	**m.** 7 lb. overweight
g. A loss of weight of 6 lb.	**n.** Walking up 10 steps

3. What number is the opposite of each of the following?

    a. $+3$ **b.** $+12$ **c.** -18 **d.** -64 **e.** -100

4. State the word that is opposite in meaning to each of the following:

a. gain	**e.** north	**i.** asset	**m.** A.D.
b. rise	**f.** east	**j.** right-hand	**n.** accelerate
c. above	**g.** forward	**k.** earnings	**o.** expand
d. up	**h.** deposit	**l.** receipt	

EXTRA FOR EXPERTS

Write the *result* for each of the following problems as a signed number:

a. Your team gains 3 yards on the first play and loses 6 yards on the next play.

b. The temperature goes up 4 degrees in 1 hour, goes up 3 degrees more in the second hour, and then goes down 5 degrees in the third hour.

2-6 *Extension of the Number Line to the Set of Integers*

STUDY GUIDE   The negative whole numbers $-1, -2, -3, \ldots$ are called *negative integers*.

The positive whole numbers $+1, +2, +3, \ldots$ are called *positive integers*. Positive integers may be written without the $+$ sign as $1, 2, 3, \ldots$.

The number 0 is neither positive nor negative.

Except for 0, any number written without a sign is a positive number.

The numbers shown on the number line below form the *set of integers* (that is, positive whole numbers, negative whole numbers, and zero).

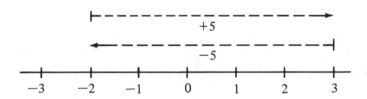

The positive numbers are represented to the right of zero, and the negative numbers to the left of zero. For this reason, movement to the right along the number line is considered movement in the positive direction. A movement to the left is said to be a movement in the negative direction. This is the reason why positive and negative numbers are called *directed numbers*.

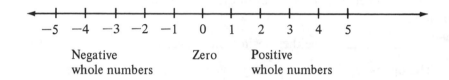

Another way to form the set of integers is to take the whole numbers $0, 1, 2, 3, \ldots$ and pair them with their opposites, $0, -1, -2, -3, \ldots$, as pictured in the diagram below.

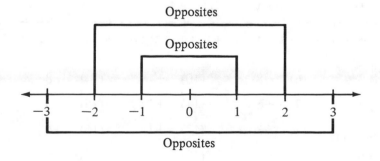

Notice that each number in a pair of opposites is the same distance from 0, but on opposite sides of 0. Zero is the opposite of 0.

1.   Numbers to the right of zero are called _____ .

2.   Numbers to the left of zero are called _____ .

3.   Other than 0, if a number has no sign in front of it, it is _____ .
Such a number is to the _____ of zero on the number line.

4.   If a number other than 0 has a minus sign in front of it, it is _____ .
This number is to the _____ of zero on the number line.

5.   On a number line, movement to the right is in the _____ direction.

6.   On a number line, movement to the left is in the _____ direction.

7.   _____ is neither positive nor negative.

8.   The opposite of a positive integer is a _____ integer.

9.   The opposite of a negative integer is a _____ integer.

10.   Two numbers the same distance from 0 but on different sides of 0 are called _____ .

11.   The set of integers includes _____ _____

_____ , _____ _____ _____ ,

and _____ .

EXERCISES

1.   Represent each of the following movements by a signed **number:**

   a.   Moving 7 units to the left   **c.**   Moving 9 units to the right

   b.   Moving 14 units to the left   **d.**   Moving 23 units to the right

2.   **a.**   Name the letter corresponding to each of the following numbers.

   b.   Name the pairs of letters that represent opposites.

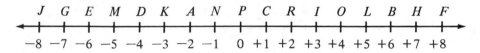

J   G   E   M   D   K   A   N   P   C   R   I   O   L   B   H   F

−8 −7 −6 −5 −4 −3 −2 −1   0 +1 +2 +3 +4 +5 +6 +7 +8

(1) +3	(6) −8	(10) −1	(14) +8
(2) −7	(7) +5	(11) +7	(15) −5
(3) +4	(8) −3	(12) +1	(16) −4
(4) +6	(9) +2	(13) −2	(17) −6
(5) 0			

3. Each of the following statements refers to moves on the number line. Sketch a number line, and name the number at which you finish.

Example: Start at -3 and move 5 units in the positive direction.

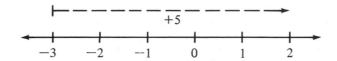

Solution: Finish at $+2$.

a. Start at $+2$ and move 5 units in the negative direction.

b. Start at $+3$ and move 2 units in the negative direction.

c. Start at -2 and move 7 units in the positive direction.

d. Start at -4 and move 3 units in the positive direction.

e. Start at -1 and move 3 units in the negative direction.

f. Start at $+4$ and move 7 units in the negative direction.

g. Start at -2, move 5 units in the positive direction, and then move 2 units in the positive direction.

h. Start at $+4$, move 3 units in the negative direction, and then move 3 units in the positive direction.

i. Start at -7, move 13 units in the negative direction, and then move 6 units in the positive direction.

4. Using a number line, express each *change* as a signed number.

a.	From 0 to $+5$	**g.**	From -2 to $+6$	**l.**	From $+5$ to $+1$
b.	From 0 to -2	**h.**	From $+3$ to 1	**m.**	From -10 to -3
c.	From -1 to $+3$	**i.**	From $+2$ to $+8$	**n.**	From $+4$ to -7
d.	From $+2$ to -2	**j.**	From -1 to -8	**o.**	From $+6$ to 0
e.	From -6 to -1	**k.**	From -6 to $+7$	**p.**	From $+4$ to -4
f.	From -3 to 0				

5. Name the signed number or numbers that is (are) described in each of the following:

a. A number 5 units to the left of 0

b. A number 6 units to the right of 0

c. Two signed numbers such that each is 6 units from 0

d. Two signed numbers such that each is 17 units from 0

e. Two signed numbers such that each is 3 units from $+18$

f. Two signed numbers such that each is 10 units from $+14$

g. Two signed numbers such that each is 5 units from $+2$

h. Two signed numbers such that each is 5 units from -2

6. For each description, give the signed number that represents the change.

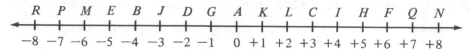

a. From A to L	**f.**	From A to E	**k.**	From N to L	**p.**	From N to R
b. From A to B	**g.**	From A to I	**l.**	From P to B	**q.**	From Q to Q
c. From A to D	**h.**	From A to R	**m.**	From R to E	**r.**	From H to B
d. From A to Q	**i.**	From A to A	**n.**	From N to A	**s.**	From I to M
e. From A to C	**j.**	From C to H	**o.**	From Q to G		

7. Name the opposite of each integer.

a. 5	**c.** -67	**e.** 11	**g.** -101	**i.** 0	**k.** -197
b. -6	**d.** 325	**f.** -12	**h.** 3000	**j.** 429	**l.** -202

EXTRA FOR EXPERTS

Use the number line below to find the number represented by each of the following:

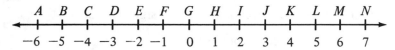

a. The point halfway between B and H

b. The point halfway between E and M

c. The point halfway between A and M

d. The point halfway between D and K

e. The point one fourth of the way from F to J

f. The point one fourth of the way from J to F

g. The point one fourth of the way from F to D

2-7.1 *Ordering Integers on the Number Line*

STUDY GUIDE On a number line containing 0 and positive numbers, the number farther to the right is the greater. Five is to the right of 3. Therefore $5 > 3$.

When the number line is extended to include negative numbers, the principle for determining whether one number is greater or smaller than another does not change.

For any two numbers represented on the number line, the number on the right is greater or the number on the left is smaller.

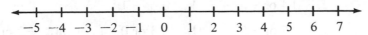

Therefore:

0 is to the right of −2,   so   0 > −2,
       or
−2 is to the left of 0,   so −2 < 0.
−2 is to the right of −4, so −2 > −4 or −4 < −2.
5 is to the right of −1,   so −1 < 5 or 5 > −1.

Thus, on the number line, numbers get *larger* as you move *to the right* and *smaller* as you move *to the left*. The number whose point is farther to the right is the greater of two numbers.

FACT FINDING

1. Of two numbers, the greater number is farther to the _____.

2. Of two numbers, the smaller number is farther to the _____.

3. Zero is to the _____ of −2.

4. As we move to the right on the number line, each number is

_____ than the number to its _____.

EXERCISES

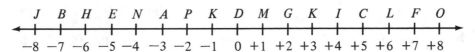

1. On the above number line, which point corresponds to the greater number?

 a. Point *A* or point *B*   **c.** Point *C* or point *H*

 b. Point *K* or point *M*   **d.** Point *P* or point *D*

2. On the above number line, which point corresponds to the smaller number?

 a. Point *G* or point *L*   **c.** Point *D* or point *N*

 b. Point *I* or point *E*   **d.** Point *F* or point *R*

3. Which number is associated with the point farther to the right on a horizontal number line?

 a. +2 or +5   **c.** 0 or +1   **e.** 0 or −3

 b. +3 or −2   **d.** −9 or +9   **f.** −5 or −4

4. Select the greater of the two given numbers.

a.	+4 or +7	**f.**	+4 or −6	**k.**	0 or +1	**p.**	0 or −6
b.	+5 or +3	**g.**	+5 or −5	**l.**	+3 or 0	**q.**	−2 or −1
c.	+9 or +6	**h.**	−3 or +2	**m.**	0 or −3	**r.**	−8 or −5
d.	+8 or +10	**i.**	0 or +6	**n.**	−2 or 0	**s.**	−3 or −4
e.	−2 or +3	**j.**	+8 or 0	**o.**	0 or −7	**t.**	−7 or −10

5. Tell whether each of the following statements is TRUE or FALSE:

a.	$3 > 2$	**f.**	$8 > -3$	**k.**	$-2 < -1$	**p.**	$-3 < -7$
b.	$4 > 6$	**g.**	$-8 < -2$	**l.**	$-4 < -1$	**q.**	$-4 < 0$
c.	$+5 > -5$	**h.**	$-6 > -3$	**m.**	$-1 < -2$	**r.**	$-2 > -6$
d.	$0 < -1$	**i.**	$-6 < 6$	**n.**	$-8 < 0$	**s.**	$1 < -3$
e.	$7 < -10$	**j.**	$-3 < 8$	**o.**	$-10 > -2$	**t.**	$0 < 9$

6. Using the symbol "<" or ">," fill in the blank to make each resulting statement true:

a.	+5 __ +1	**g.**	13 __ −13	**m.**	9 __ −1	**s.**	+99 __ −2
b.	+8 __ +11	**h.**	−1 __ 1	**n.**	2 __ −5	**t.**	−7 __ −3
c.	+1 __ +3	**i.**	0 __ −3	**o.**	10 __ −10	**u.**	−5 __ −6
d.	6 __ 0	**j.**	+3 __ −2	**p.**	−10 __ 10	**v.**	−2 __ −11
e.	0 __ 6	**k.**	−2 __ +1	**q.**	+2 __ −200	**w.**	−5 __ −6
f.	−3 __ 0	**l.**	−1 __ 9	**r.**	−5000 __ +3	**x.**	−8 __ −1

7. Name the following numbers in order of size (smallest first):

a.	6, 9, 4	**g.**	−6, 2, −4	**m.**	−15, −17, −13
b.	8, 5, −10	**h.**	6, −2, 4	**n.**	0, −21, 21
c.	7, 0, −9	**i.**	−3, +3, −4	**o.**	−50, −49, −51
d.	0, −2, −8	**j.**	−8, 3, −2	**p.**	6, −4, 3, −1
e.	1, −5, −11	**k.**	−10, 10, −1	**q.**	1, −1, 2, −2
f.	−5, −10, −15	**l.**	−4, −8, −1	**r.**	7, −3, −5, 9

EXTRA FOR EXPERTS

Write an inequality to compare the three numbers in each of the following.
Example: −3, 1, and −4

Answer: $-4 < -3 < 1$

a.	0, −4, and 4	**c.**	−3, −5, and 0	**e.**	−4, 6, and −2
b.	−1, 2, and −6	**d.**	3, −5, and −6	**f.**	−2, −1, and −6

2-7.2 *Ordering of Positive and Negative Rational Numbers*

STUDY GUIDE If two numbers are named by fractions, the greater fraction is farther to the right on the number line or ruler. Also, if two fractions have the same denominator, the fraction with the greater numerator has the greater value.

Recall that two fractions naming the same number are called *equivalent fractions*. To change a fraction into an equivalent fraction with another denominator, we multiply the numerator and the denominator by the same nonzero number.

To make a comparison between two fractions, the fractions must have the same denominator. This denominator is called a *common denominator*. The fractions $\frac{1}{2}$ and $\frac{3}{8}$ can each be changed to equivalent fractions with the same denominator:

$$\frac{1}{2} = \frac{4}{8} = \frac{8}{16} = \frac{12}{24}, \text{ and so on.}$$

$$\frac{3}{8} = \frac{6}{16} = \frac{9}{24}, \text{ and so on.}$$

The numbers 8, 16, and 24 are common denominators.

The *lowest common denominator* (abbreviated LCD) is the smallest number that is a common denominator.

Eight is the smallest of the common denominators for the fractions given above, and 8 is called the lowest common denominator.

FACT FINDING

1. On a number line the smaller of two fractions is located to the

_____ of the greater fraction.

2. Two fractions that name the same number are called _____ fractions.

3. To obtain an equivalent fraction, we multiply the _____

and the _____ by the same nonzero number.

4. If two fractions having the same denominator are compared, the greater

fraction has _____ _____ _____ .

5. To compare two fractions having different denominators, they must be

changed to _____ fractions with the same _____ .

6. The smallest number that is a common denominator is called the

_____ _____ _____ .

MODEL PROBLEM

Fill in the blank with the symbol "<" or ">" to make this statement true: $\dfrac{5}{12} - \dfrac{1}{2}$.

Solution: (1) Find the LCD.
Since

$$\left(\dfrac{1}{2}\right) = \dfrac{2}{4} = \dfrac{3}{6} = \dfrac{4}{8} = \dfrac{5}{10} = \left(\dfrac{6}{12}\right) = \dfrac{7}{14}, \text{ and so on,}$$

and

$$\dfrac{5}{12} = \dfrac{10}{24} = \dfrac{15}{36}, \text{ and so on,}$$

therefore the LCD is 12.

(2) Compare $\dfrac{5}{12}$ and $\dfrac{6}{12}$. Since the fractions $\dfrac{5}{12}$ and $\dfrac{6}{12}$ have the same denominator, we can compare the numerators. Since the numerator 6 in $\dfrac{6}{12}$ is greater than the numerator 5 in $\dfrac{5}{12}$, then $\dfrac{6}{12}$ is the greater fraction.

(3) Therefore

$$\dfrac{6}{12} > \dfrac{5}{12} \qquad \text{or} \qquad \dfrac{5}{12} < \dfrac{6}{12}.$$

So

$$\dfrac{1}{2} > \dfrac{5}{12} \qquad \text{or} \qquad \dfrac{5}{12} < \dfrac{1}{2}.$$

(4) Answer: $\dfrac{5}{12} < \dfrac{1}{2}$.

EXERCISES

1. Find the lowest common denominator for each group of examples:

a. $\dfrac{1}{2}$ and $\dfrac{5}{12}$ 　　　 c. $\dfrac{7}{8}$ and $\dfrac{1}{2}$ 　　　 e. $\dfrac{-4}{5}$ and $\dfrac{-5}{15}$

b. $\dfrac{-2}{3}$ and $\dfrac{5}{9}$ 　　　 d. $\dfrac{3}{10}$ and $\dfrac{-7}{30}$ 　　　 f. $\dfrac{5}{12}, \dfrac{2}{3},$ and $\dfrac{1}{24}$

2. Supply the missing number to make each statement true.

a. $\dfrac{3}{5} = \dfrac{}{10}$ 　　 d. $\dfrac{2}{3} = \dfrac{}{9}$ 　　 g. $\dfrac{1}{6} = \dfrac{}{30}$ 　　 j. $\dfrac{3}{4} = \dfrac{9}{}$

b. $\dfrac{5}{6} = \dfrac{}{48}$ 　　 e. $\dfrac{11}{25} = \dfrac{44}{}$ 　　 h. $\dfrac{3}{2} = \dfrac{}{8}$ 　　 k. $\dfrac{5}{8} = \dfrac{15}{}$

c. $\dfrac{7}{9} = \dfrac{63}{}$ 　　 f. $\dfrac{2}{3} = \dfrac{}{18}$ 　　 i. $\dfrac{2}{3} = \dfrac{8}{}$ 　　 l. $\dfrac{5}{6} = \dfrac{20}{}$

3. Fill in the blank with the symbol "<," "=," or ">" to make each statement true. (*Hint:* Change mixed numbers to improper fractions before comparing.)

a. $\dfrac{+1}{8} \underline{\quad} \dfrac{+5}{8}$

b. $\dfrac{-1}{4} \underline{\quad} \dfrac{+3}{4}$

c. $\dfrac{-1}{8} \underline{\quad} \dfrac{-7}{8}$

d. $-1\dfrac{1}{4} \underline{\quad} \dfrac{-3}{4}$

e. $\dfrac{+5}{9} \underline{\quad} \dfrac{-5}{9}$

f. $-1\dfrac{1}{8} \underline{\quad} \dfrac{-7}{8}$

g. $\dfrac{+1}{4} \underline{\quad} \dfrac{-3}{4}$

h. $\dfrac{3}{16} \underline{\quad} \dfrac{-9}{16}$

i. $-2\dfrac{2}{3} \underline{\quad} -3\dfrac{1}{3}$

j. $\dfrac{3}{4} \underline{\quad} \dfrac{-3}{4}$

k. $\dfrac{+2}{3} \underline{\quad} \dfrac{+1}{3}$

l. $\dfrac{-2}{3} \underline{\quad} \dfrac{-1}{3}$

m. $-1\dfrac{1}{3} \underline{\quad} \dfrac{-4}{3}$

n. $+3\dfrac{7}{8} \underline{\quad} +2\dfrac{7}{8}$

o. $-3\dfrac{3}{8} \underline{\quad} -3\dfrac{5}{8}$

p. $\dfrac{1}{4} \underline{\quad} 0$

q. $\dfrac{-1}{4} \underline{\quad} 0$

r. $-2\dfrac{1}{2} \underline{\quad} 1$

s. $-2 \underline{\quad} -1\dfrac{1}{2}$

t. $+4\dfrac{1}{2} \underline{\quad} 5$

u. $-4\dfrac{1}{2} \underline{\quad} -5$

4. Fill in the blank with the symbol "<," "=," or ">" to make each statement true.

a. $\dfrac{3}{4} \underline{\quad} \dfrac{7}{8}$

b. $\dfrac{1}{4} \underline{\quad} \dfrac{7}{16}$

c. $\dfrac{1}{4} \underline{\quad} \dfrac{1}{12}$

d. $\dfrac{-3}{8} \underline{\quad} \dfrac{3}{4}$

e. $\dfrac{-3}{8} \underline{\quad} \dfrac{-3}{4}$

f. $\dfrac{10}{4} \underline{\quad} \dfrac{5}{2}$

g. $\dfrac{1}{2} \underline{\quad} \dfrac{11}{12}$

h. $\dfrac{4}{5} \underline{\quad} \dfrac{3}{10}$

i. $\dfrac{12}{24} \underline{\quad} \dfrac{6}{12}$

j. $\dfrac{-1}{4} \underline{\quad} \dfrac{-1}{12}$

k. $\dfrac{3}{6} \underline{\quad} \dfrac{4}{12}$

l. $\dfrac{-4}{3} \underline{\quad} \dfrac{-5}{6}$

m. $\dfrac{-5}{2} \underline{\quad} \dfrac{-3}{4}$

n. $-1\dfrac{1}{4} \underline{\quad} -1\dfrac{1}{2}$

o. $-2\dfrac{11}{12} \underline{\quad} -2\dfrac{2}{3}$

p. $-3\dfrac{7}{8} \underline{\quad} -3\dfrac{1}{4}$

EXTRA FOR EXPERTS

1. In the number line below, *W, S, A, T, R,* and *M* are signed numbers. Referring to this number line, tell which statements are TRUE and which are FALSE.

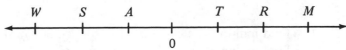

a. *A* is a negative number.

b. *T* is a positive number.

c. *W* is greater than zero.

d. *S* is a negative number.

e. *T* is greater than zero.

f. *W* is not a positive number.

g. $T > A$.

h. $R < M$.

i. $W > 0$.

j. $S > W$.

k. $W > A$.

l. $O > S$.

2. Name the following numbers in order of size (smallest first):

   a. -14, $+9$, $+19$, -15, -22, $+7$, $+10$, -6, $+2$, -30

   b. $\dfrac{-4}{2}$, $\dfrac{-3}{2}$, $\dfrac{-1}{2}$, $\dfrac{-5}{2}$, -1, -3

2-8 *Comparing Rational Numbers*

STUDY GUIDE The numbers 3, 6, 9, 15, and so on are called *multiples* of 3. Multiples are a set of numbers that are found by starting with any given number and multiplying by 1, 2, 3, ... If two sets of multiples are written, such as

   multiples of 3:   3, 6, 9, 12, (15), 18, 21, 24, 27, 30, . . .
   multiples of 5:   5, 10, (15), 20, 25, 30, 35, . . .

then the multiples that are found in both sets are called *common multiples*. The smallest common multiple is called the *least common multiple* (abbreviated as LCM). In the two sets of multiples given above, 15 and 30 are common multiples. Fifteen is the lowest common multiple.

   Common multiples can be used to find the lowest common denominator (LCD). The lowest common denominator is the least common multiple of the different denominators.

FACT FINDING

1. _____ are a set of numbers that are found by starting with any given number and _____ by 1, 2, 3, ...

2. _____ _____ are multiples that are found in both sets.

3. The smallest common multiple is the _____ common multiple.

4. The _____ common denominator is the least common _____ of the different denominators.

MODEL PROBLEM

   Write the symbol "<" or ">" in the blank so that the statement will be true: $\dfrac{3}{5} \underline{\hspace{1cm}} \dfrac{4}{7}$.

Solution: (1)   Find multiples of 5: 5, 10, 15, 20, 25, 30, (35), ...
                  Find multiples of 7: 7, 14, 21, 28, (35), 42, ...

           (2)   Find the least common multiple of the above two sets of multiples:

$$\text{LCM} = 35.$$

           (3)   Find the LCD:

$$\text{LCD} = \text{LCM} = 35.$$

(4) Change each fraction to an equivalent fraction with a denominator of 35:

$$\frac{3}{5} = \frac{3 \cdot 7}{5 \cdot 7} = \frac{21}{35} \qquad \text{and} \qquad \frac{4}{7} = \frac{4 \cdot 5}{7 \cdot 5} = \frac{20}{35}.$$

(5) Since $\frac{21}{35}$ and $\frac{20}{35}$ have the same denominator, compare the numerators.

Numerator 21 in $\frac{21}{35} >$ numerator 20 in $\frac{20}{35}$.

Therefore

$$\frac{21}{35} > \frac{20}{35}.$$

(6) Answer: $\frac{3}{5} > \frac{4}{7}$.

EXERCISES

1. Find two common multiples of each pair of numbers:

   a. 4, 5   **c.** 3, 9   **e.** 7, 3   **g.** 10, 4

   b. 6, 10   **d.** 4, 6   **f.** 9, 4   **h.** 20, 15

2. Use the method of LCM to find the LCD for each pair of fractions.

   a. $\frac{2}{3}, \frac{3}{4}$ **d.** $\frac{1}{3}, \frac{-3}{7}$ **g.** $\frac{5}{8}, \frac{5}{7}$ **j.** $\frac{3}{4}, \frac{1}{10}$

   b. $\frac{1}{5}, \frac{1}{4}$ **e.** $\frac{-4}{9}, \frac{-3}{8}$ **h.** $\frac{2}{3}, \frac{-4}{9}$ **k.** $\frac{7}{10}, \frac{9}{11}$

   c. $\frac{5}{6}, \frac{2}{5}$ **f.** $\frac{1}{6}, \frac{1}{10}$ **i.** $\frac{9}{10}, \frac{1}{6}$ **l.** $\frac{2}{7}, \frac{-1}{4}$

3. Express each set of fractions so that they are equivalent fractions with the least common denominator.

   a. $\frac{1}{3}, \frac{3}{5}$ **c.** $\frac{5}{6}, \frac{3}{14}$ **e.** $\frac{3}{8}, \frac{-13}{24}$ **g.** $\frac{3}{4}, \frac{17}{10}$

   b. $\frac{11}{6}, \frac{7}{10}$ **d.** $\frac{-3}{4}, \frac{2}{7}$ **f.** $\frac{-3}{20}, \frac{-7}{30}$ **h.** $\frac{5}{18}, \frac{13}{4}$

4. Write the symbol "<" or ">" in the blank so that each statement will be true.

a. $\dfrac{1}{2} \underline{\hspace{1em}} \dfrac{4}{7}$ g. $\dfrac{-2}{7} \underline{\hspace{1em}} \dfrac{-1}{4}$ m. $\dfrac{-4}{5} \underline{\hspace{1em}} \dfrac{-7}{9}$ s. $1\dfrac{1}{3} \underline{\hspace{1em}} 1\dfrac{1}{5}$

b. $\dfrac{5}{6} \underline{\hspace{1em}} \dfrac{7}{9}$ h. $\dfrac{-5}{8} \underline{\hspace{1em}} \dfrac{-11}{15}$ n. $\dfrac{31}{3} \underline{\hspace{1em}} \dfrac{51}{5}$ t. $2\dfrac{2}{3} \underline{\hspace{1em}} 2\dfrac{3}{4}$

c. $\dfrac{5}{8} \underline{\hspace{1em}} \dfrac{13}{20}$ i. $\dfrac{4}{7} \underline{\hspace{1em}} \dfrac{4}{5}$ o. $\dfrac{16}{38} \underline{\hspace{1em}} \dfrac{3}{7}$ u. $2\dfrac{3}{4} \underline{\hspace{1em}} 2\dfrac{5}{6}$

d. $\dfrac{5}{6} \underline{\hspace{1em}} \dfrac{5}{7}$ j. $\dfrac{7}{10} \underline{\hspace{1em}} \dfrac{6}{11}$ p. $\dfrac{6}{9} \underline{\hspace{1em}} \dfrac{11}{15}$ v. $7\dfrac{1}{4} \underline{\hspace{1em}} 7\dfrac{1}{8}$

e. $\dfrac{-5}{8} \underline{\hspace{1em}} \dfrac{-10}{17}$ k. $\dfrac{10}{11} \underline{\hspace{1em}} \dfrac{3}{4}$ q. $\dfrac{-9}{7} \underline{\hspace{1em}} \dfrac{-5}{6}$ w. $\dfrac{-15}{12} \underline{\hspace{1em}} \dfrac{-5}{8}$

f. $\dfrac{-4}{5} \underline{\hspace{1em}} \dfrac{-3}{8}$ l. $\dfrac{9}{10} \underline{\hspace{1em}} \dfrac{11}{12}$ r. $\dfrac{-2}{3} \underline{\hspace{1em}} \dfrac{-4}{7}$ x. $5\dfrac{1}{40} \underline{\hspace{1em}} 5\dfrac{1}{50}$

EXTRA FOR EXPERTS

1. Find the number halfway between each pair of fractions.

a. $\dfrac{1}{6}$ and $\dfrac{2}{6}$ b. $\dfrac{2}{3}$ and $\dfrac{3}{3}$

2. Large candy bars weigh $8\dfrac{3}{4}$ ounces each, and small candy bars weigh $\dfrac{7}{8}$ of an ounce each.

a. Find the weight of 3 large candy bars and 24 small candy bars.

b. By how many ounces does the heavier candy bar exceed the lighter?

2-9.1 *Comparing More Than Two Rational Numbers*

STUDY GUIDE Recall the Study Guide from Section 2-8.
To find the multiples of a number, multiply the number by 1, 2, 3, 4, ... When a multiple of one whole number is also a multiple of another whole number, it is called a *common multiple* of the two numbers. The smallest number that is a multiple of two or more numbers is called the *least common multiple*.

To compare two or more rational numbers, the denominators have to be the same number. The common denominator used to rename each fraction should be the *lowest common denominator*. The lowest common denominator is the *least common multiple* of the different denominators.

FACT FINDING

1. A multiple of a number is found by _____ that number by any counting number, that is, 1, 2, 3, ...

2. When a multiple of one whole number is also a multiple of another whole number, it is called a _____ _____ of the two numbers.

3. The smallest number that is a multiple of two or more numbers is called the _____ _____ _____ .

4. The least common denominator of two or more numbers is defined as the _____ common multiple of the _____ _____ .

5. Renaming fractions into equivalent fractions means multiplying each fraction by a fraction whose value is _____ .

MODEL PROBLEM

Arrange the fractions $\frac{1}{3}, \frac{1}{6}, \frac{3}{4}$ in order from smallest to largest.

Solution: (1) Find the least common denominator by finding:

      a. Multiples of 3: 3, 6, 9, ⑫, 15, 18, 21, 24, ...

         Multiples of 6: 6, ⑫, 18, 24, 30, ...

         Multiples of 4: 4, 8, ⑫, 16, 20, 24, 28, ...

      b. The least common multiple of all denominators = 12. Therefore the LCD = 12.

(2) Rename each fraction so that it is equivalent to the original fraction but has the LCD as its denominator:

$$\frac{1}{3} = \frac{1 \cdot 4}{3 \cdot 4} = \frac{4}{12}; \quad \frac{1}{6} = \frac{1 \cdot 2}{6 \cdot 2} = \frac{2}{12}; \quad \frac{3}{4} = \frac{3 \cdot 3}{4 \cdot 3} = \frac{9}{12}.$$

(3) Since the three fractions all have the same denominator, compare the numerators. Numerator 2 in $\frac{2}{12} <$ numerator 4 in $\frac{4}{12} <$ numerator 9 in $\frac{9}{12}$.

(4) Answer: $\frac{2}{12} < \frac{4}{12} < \frac{9}{12}$ or $\frac{1}{6} < \frac{1}{3} < \frac{3}{4}$.

EXERCISES

1. Find two common multiples for each set of three numbers.

    a. 6, 9, 12   **c.** 8, 9, 6   **e.** 4, 5, 6

    b. 4, 9, 15   **d.** 7, 4, 8   **f.** 8, 10, 12

2. Find the LCD for each group of fractions.

a. $\dfrac{3}{5}, \dfrac{2}{3}, \dfrac{7}{10}$ d. $\dfrac{-2}{7}, \dfrac{-3}{4}, \dfrac{-15}{24}$ g. $\dfrac{3}{10}, \dfrac{9}{50}, \dfrac{7}{25}$

b. $\dfrac{7}{8}, \dfrac{11}{12}, \dfrac{2}{3}$ e. $\dfrac{-3}{5}, \dfrac{-4}{10}, \dfrac{-2}{3}$ h. $\dfrac{-3}{4}, \dfrac{7}{18}, \dfrac{-30}{36}$

c. $\dfrac{3}{5}, \dfrac{2}{3}, \dfrac{7}{15}$ f. $\dfrac{1}{7}, \dfrac{3}{14}, \dfrac{5}{6}$ i. $\dfrac{-3}{5}, \dfrac{-7}{15}, \dfrac{-1}{4}, \dfrac{-1}{3}$

3. Arrange the fractions in each group in order of value from least to greatest.

a. $\dfrac{3}{5}, \dfrac{5}{2}, \dfrac{9}{10}$ h. $\dfrac{3}{4}, \dfrac{5}{6}, \dfrac{1}{8}$ o. $\dfrac{5}{9}, \dfrac{-7}{13}, \dfrac{-6}{11}$

b. $\dfrac{2}{3}, \dfrac{4}{5}, \dfrac{1}{2}$ i. $\dfrac{4}{4}, \dfrac{13}{16}, \dfrac{5}{8}$ p. $\dfrac{-1}{3}, \dfrac{-5}{12}, \dfrac{-5}{14}$

c. $\dfrac{-1}{2}, \dfrac{-2}{3}, \dfrac{-2}{5}$ j. $\dfrac{15}{16}, \dfrac{2}{2}, \dfrac{7}{8}$ q. $\dfrac{1}{3}, \dfrac{2}{5}, \dfrac{3}{7}$

d. $\dfrac{-3}{4}, \dfrac{-7}{18}, \dfrac{-30}{36}$ k. $\dfrac{-13}{16}, \dfrac{-3}{8}, \dfrac{-3}{4}$ r. $-1\dfrac{1}{2}, -1\dfrac{1}{3}, -1\dfrac{5}{12}$

e. $\dfrac{-2}{3}, \dfrac{-4}{5}, \dfrac{-1}{2}$ l. $\dfrac{11}{12}, \dfrac{3}{2}, \dfrac{7}{6}$ s. $4\dfrac{2}{3}, 4\dfrac{3}{5}, 4\dfrac{1}{2}$

f. $\dfrac{4}{9}, \dfrac{1}{6}, \dfrac{5}{2}$ m. $\dfrac{-3}{4}, \dfrac{-1}{2}, \dfrac{-3}{8}$ t. $-2\dfrac{2}{3}, -2\dfrac{2}{7}, -2\dfrac{1}{6}$

g. $\dfrac{-7}{8}, \dfrac{-1}{6}, \dfrac{-7}{4}$ n. $\dfrac{1}{3}, \dfrac{3}{8}, \dfrac{2}{7}$ u. $3\dfrac{1}{8}, 3\dfrac{5}{12}, 3\dfrac{5}{6}$

EXTRA FOR EXPERTS

Double the denominator of a fraction, but do not change the numerator. What change, if any, takes place in the fraction? Explain.

2-9.2 Comparing Fractions Having the Same Numerator but Different Denominators

STUDY GUIDE To compare $\dfrac{3}{4}$ and $\dfrac{3}{8}$, it is not necessary to find a common denominator. The numerator of a fraction indicates how many parts of the whole are being considered. The denominator of a fraction indicates into how many parts the whole is divided. The smaller of the two denominators divides the whole into fewer parts, so that each part is larger.

Therefore, if two fractions have the *same numerator*, the fraction that has the smaller denominator names the fraction that has the greater value. Thus $\dfrac{3}{4} > \dfrac{3}{8}$ because 3 out of 4 equal parts of the whole is larger than 3 out of 8 equal parts of the whole.

FACT FINDING

1. If two fractions have the same denominator, the greater fraction is the one that has the _____ _____ .

2. If two fractions have the same numerator, the greater fraction is the one that has the _____ _____ .

MODEL PROBLEM

Arrange the following in order from smallest to largest: $\dfrac{6}{9}, \dfrac{6}{5}, \dfrac{6}{11}$.

Solution: (1) Since the numerators are the same, the fraction with the largest denominator has the smallest value. Eleven is the largest denominator.

(2) Arrange the denominators in order of size: 11, 9, 5.

(3) Answer: $\dfrac{6}{11} < \dfrac{6}{9} < \dfrac{6}{5}$.

EXERCISES

1. Arrange each of the following in order from smallest to largest:

   a. $\dfrac{5}{3}, \dfrac{5}{6}$ c. $\dfrac{1}{3}, \dfrac{1}{4}$ e. $\dfrac{3}{4}, \dfrac{3}{5}$ g. $\dfrac{4}{7}, \dfrac{4}{10}$

   b. $\dfrac{2}{5}, \dfrac{2}{7}$ d. $\dfrac{9}{11}, \dfrac{9}{8}$ f. $\dfrac{1}{6}, \dfrac{1}{4}$ h. $\dfrac{7}{8}, \dfrac{7}{9}$

   i. $\dfrac{7}{4}, \dfrac{7}{2}, \dfrac{7}{8}$ j. $\dfrac{9}{2}, \dfrac{9}{4}, \dfrac{9}{8}$ k. $\dfrac{5}{3}, \dfrac{5}{9}, \dfrac{5}{7}$

2. Which time is greater: $\dfrac{1}{7}$ of an hour or $\dfrac{1}{5}$ of an hour?

3. Which is narrower: a screw $\dfrac{7}{8}$ of an inch wide or a screw $\dfrac{7}{16}$ of an inch wide?

4. Which is the greater distance: $\dfrac{17}{10}$ of a mile or $\dfrac{17}{5}$ of a mile?

5. Which portion of pizza is smaller: $\dfrac{4}{7}$ of a pie or $\dfrac{4}{9}$ of a pie?

EXTRA FOR EXPERTS

When the denominator of a fraction is decreased and the numerator does not change, is the value of the fraction increased or decreased?

2-10 *Adding Integers Using the Number Line*

To use the number line to find the *sum of two integers,* you should recall that:

1. A positive integer can be represented by a "move to the right."
2. A negative integer can be represented by a "move to the left."

Examples: +3 represents a move of 3 units to the right.
−2 represents a move of 2 units to the left.

Moves on a number line can be used to find the sum of two integers. *Example:* Find (+3) + (−2).

(1) Starting at 0 on the number line, move the number of units indicated by the first integer.

1. Move from 0 to +3.

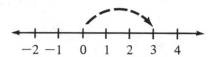

(2) Starting at the result of step (1), move the number of units indicated by the second integer.

2. From +3 move −2 units.

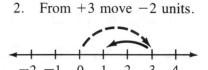

(3) The number you arrive at is the sum.

3. Then +1 is the sum.

Answer: (+3) + (−2) = +1 (or 1).

1. A move to the right on the number line represents a _____ integer.

2. A move to the left on the number line represents a _____ integer.

3. When adding a set of numbers on a number line, the starting point for the procedure is _____.

Use the number line to find the following sums:

a.   (+3) + (+2).

    Solution: (1) Move from 0 to +3.

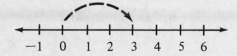

                (2) From +3 move +2 units.

                (3) Then +5 is the sum.

                (4) Answer: (+3) + (+2) = +5.

b.   (−3) + (−2).

    Solution: (1) Move from 0 to −3.

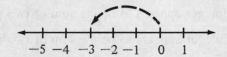

                (2) From −3 move −2 units.

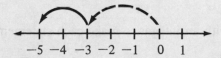

                (3) Then −5 is the sum.

                (4) Answer: (−3) + (−2) = −5.

c.   (+2) + (−5).

    Solution: (1) Move from 0 to +2.

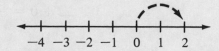

                (2) From +2 move −5 units.

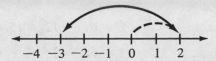

                (3) Then −3 is the sum.

                (4) Answer: (+2) + (−5) = −3.

1. Make a diagram picturing each sum. In each case, also write the sum.

   a. $(+2) + (+5)$ **e.** $3 + 5$ **i.** $(+8) + (-8)$

   b. $(-2) + (-5)$ **f.** $(+6) + (-2)$ **j.** $(-7) + (+7)$

   c. $(-6) + (-2)$ **g.** $(-6) + (-2)$ **k.** $(-4) + (-3)$

   d. $4 + 2$ **h.** $(-5) + (+4)$ **l.** $(-16) + 5$

2. Use a number line to find each of the following sums:

   a. $4 + 2$ **e.** $(-4) + (-2)$ **i.** $5 + (-8)$ **m.** $-6 + 4$

   b. $1 + 6$ **f.** $(-6) + (-4)$ **j.** $(-8) + 5$ **n.** $(-7) + (-4)$

   c. $(+5) + (-3)$ **g.** $10 + (-7)$ **k.** $2 + (-9)$ **o.** $(-2) + (-2)$

   d. $(-3) + (+5)$ **h.** $-1 + 4$ **l.** $5 + (-3)$ **p.** $(-2) + (-8)$

   q. $(-3) + (+1) + (-5)$ **s.** $-6 + (-2) + (-5)$

   r. $(+4) + (-8) + (+6)$ **t.** $(-8) + (+13) + (-20) + (+9)$

3. **a.** What type of integer appears to be the sum of two positive integers?

   b. What type of integer appears to be the sum of two negative integers?

   c. What type of integer results when we find the sum of a positive and a negative integer?

4. Fill in the blank with an integer to make each statement true.

   a. $-4 + (-3) =$ __   **g.** $7 +$ __ $= -3$

   b. __ $+ (-3) = -8$ **h.** $-7 + 0 =$ __

   c. $-2 + 9 =$ __   **i.** $-2 +$ __ $= 0$

   d. $-8 + (-1) =$ __   **j.** $-10 +$ __ $= -3$

   e. $-6 +$ __ $= -1$ **k.** $-7 +$ __ $= 8$

   f. __ $+ 3 = -5$ **l.** __ $+ 0 = -3$

5. If $9 - 5$ means $9 + (-5)$, fill in the blank with an integer to make each statement true.

   a. $6 - 7 = 6 +$ __   **d.** $-12 - 15 = -12 +$ __

   b. $5 - 12 = 5 +$ __   **e.** $-3 - 2 - 4 = -3 +$ __ $+$ __

   c. $-3 - 5 = -3 +$ __

1. In March, Tom made deposits of $30 and $20. Later, he withdrew $40 and $30. Find the change in his balance due to these deposits and withdrawals, using signed numbers.

2. If the temperature was $-12°$ at 6 A.M. and $-2°$ at noon, what was the change in temperature?

3. Lucy's kite was flying at a height of 800 m. She lowered it 200 m, raised it 500 m, lowered it 300 m, and raised it 700 m. At what height is the kite flying now? (Explain your answer using signed numbers.)

4. Find the sum in each part, and state the relationship between the sums.

    a. $(+1) + (+6)$ **b.** $(-5) + (-4)$ **c.** $(+7) + (-3)$
        $(+6) + (+1)$ $(-4) + (-5)$ $(-3) + (+7)$

5. From the answers obtained to Exercise 4, what property of addition appears to hold true for integers?

2-11.1 *Rules for Addition of Signed Numbers*

STUDY GUIDE

A very large number line would be required to represent the sum of $+905$ and -963. Fortunately there is another method of computing the sum of two signed numbers. The rules for addition of signed numbers require the introduction of a new term, the absolute value of a number. The *absolute value* of an integer is defined as the distance of that integer from 0.

Therefore the absolute value of $+5$ is 5 because $+5$ is 5 units from 0, and the absolute value of -6 is 6 because -6 is 6 units from 0.

Using the term "absolute value," we can now state the following rules for adding signed numbers:

I. To add two positive integers:

Rule	Example
	$(+15) + (+27)$
1. Find the absolute value of each number.	Absolute value of $+15$ is 15. Absolute value of $+27$ is 27.
2. Find the sum of the absolute values.	Sum of absolute values $= 42$.
3. Find the common sign.	Common sign is $+$.
4. Place the common sign before the sum.	Answer: $+42$

II. To add two negative integers:

Rule	Example
	$(-27) + (-35)$
1. Find the absolute value of each number.	Absolute value of -27 is 27. Absolute value of -35 is 35.
2. Find the sum of the absolute values.	Sum of absolute values = 62.
3. Find the common sign.	Common sign is $-$.
4. Place the common sign before the sum.	Answer: -62

III. To add a positive and a negative integer:

Rule	Example
	(a) $(+46) + (-31)$
1. Find the absolute value of each number.	Absolute value of $+46$ is 46. Absolute value of -31 is 31.
2. Find the difference between the absolute values.	Difference between the absolute values = 15.
3. Find the sign of the number that has the greater absolute value.	Sign of the number that has the greater absolute value is $+$.
4. Place the sign from step 3 before the difference in step 2.	Answer: $+15$

Rule	Example
	(b) $(+18) + (-25)$
1. Find the absolute value of each number.	Absolute value of $+18$ is 18. Absolute value of -25 is 25.
2. Find the difference between the absolute values.	Difference between the absolute values = 7.
3. Find the sign of the number that has the greater absolute value.	Sign of the number that has the greater absolute value is $-$.
4. Place the sign from step 3 before the difference in step 2.	Answer: -7

FACT FINDING

1. When adding two numbers having the same sign, find the sum of their

_____ _____ . Place the _____ _____ in front of the sum.

2. When adding two numbers having unlike signs, find the _____

between their absolute values. Place the _____ _____

in front of the _____ .

3. The distance an integer is from 0 on the number line is called the

_____ _____ of the integer.

1. Find the absolute value of each of the following integers:

    a. -3 **b.** $+3$ **c.** -10 **d.** -167 **e.** $+91$

2. Which is greater?

    a. The absolute value of $+10$ or the absolute value of -11

    b. The absolute value of -3 or the absolute value of -8

    c. The absolute value of $+9$ or the absolute value of -10

    d. The absolute value of -4 or the absolute value of $+5$

3. State whether each of the following is TRUE or FALSE:

    a. The absolute value of -7 is greater than the absolute value of -2.

    b. The absolute value of -1 is less than the absolute value of -3.

    c. The absolute value of -10 is greater than the absolute value of $+10$.

    d. The sum of -3 and the absolute value of -3 is 0.

    e. The sum of the absolute value of -3 and the absolute value of $+3$ is 0.

4. Complete each of the following statements so that the resulting statement will be true:

    a. The sum of two positive numbers is a _____ number.

    b. The sum of two negative numbers is a _____ number.

    c. The absolute value of a positive number is a _____ number.

    d. The absolute value of a negative number is a _____ number.

5. Answer each of the following:

    a. When a positive number is added to a negative number, what determines the sign of the sum?

    b. When two positive numbers are added, what is the sign of the sum?

    c. When two negative numbers are added, what is the sign of the sum?

6. State the following rules:

    a. The "rule for addition of signed numbers with like signs."

    b. The "rule for addition of signed numbers with unlike signs."

7. Find the sum for each of the following:

a.	$(+7) + (+19)$	**h.**	$-11 + (-12)$	**o.**	$-34 + 13$	**u.**	$9 + (-9)$
b.	$+70 + 40$	**i.**	$(20) + (-6)$	**p.**	$9 + (-17)$	**v.**	$43 + (-43)$
c.	$26 + 3$	**j.**	$(-3) + 25$	**q.**	$10 + (-21)$	**w.**	$(-18) + 18$
d.	$150 + 61$	**k.**	$32 + (-16)$	**r.**	$34 + (-23)$	**x.**	$-80 + 80$
e.	$(-1) + (-11)$	**l.**	$10 + (-3)$	**s.**	$(-42) + 81$	**y.**	$-11 + 0$
f.	$(-34) + (-12)$	**m.**	$-9 + 12$	**t.**	$25 + (-19)$	**z.**	$0 + 10$
g.	$-45 + (-16)$	**n.**	$-15 + 40$				

8. Find the sum for each of the following. First study the example.

Example: $(+9) + (-4) + (+2) + (-3)$.

Solution: (1) Add the positive numbers; their sum is positive:
$$(+9) + (+2) = (+11).$$

(2) Add the negative numbers; their sum is negative:
$$(-4) + (-3) = (-7).$$

(3) Add the resulting sums, using the rule for addition of numbers with unlike signs:
$$(+11) + (-7) = +4 \text{ or } 4.$$

a. $(-8) + (-4) + (+6)$

b. $(+48) + (-12) + (-24)$

c. $(-24) + (-45) + 17 + 52$

d. $45 + (-22) + 37 + (-23) + 19$

e. $(+18) + (-6) + 13 + (-2)$

f. $(-13) + 6 + (-5) + (-2)$

g. $(-12) + 5 + 6 + (-8)$

h. $-53 + 26 + 42 + (-31) + 11$

i. $45 + (-89) + 64 + (-26)$

j. $8 + (-3) + 2 + (-9) + 1$

k. $16 + (-7) + (-12) + 22 + (-15)$

l. $(-27) + 32 + (-125) + (-56) + 78$

m. $-6 + 10 + (-7) + (-8) + 14$

n. $(-15) + 12 + (-8) + (-7) + 17$

o. $42 + (-60) + 18 + (-74)$

p. $25 + (-17) + 21 + (-34)$

q. $35 + (-70) + 24 + 30$

r. $-9 + 7 + (-5) + 8 + 27 + (-31) + 15 + 12$

9. Solve each problem, using addition of signed numbers.

    a. The enrollment in a high school shows the following increases and decreases over a 5-month period: $+100$, -200, $+150$, -70, $+30$. Find the net increase or decrease.

    b. The following changes in snow level were observed: snowed 45 cm, snowed 9 cm, melted 6 cm, melted 10 cm, melted 3 cm. Find the net increase or decrease in snow level.

2-11.2 *Opposites; Additive Inverses*

STUDY GUIDE The opposite of a number was defined as a number that was the same distance away from 0 but on the opposite side of 0. Therefore the opposite of 8 is -8, and the opposite of -8 is 8. If two opposites are added, their sum is 0. For example, $(+8) + (-8) = 0$.

The opposite of a number is also called the *additive inverse* of the number. Therefore:

or
$$-11 \text{ is the additive inverse of } +11$$
$$+11 \text{ is the additive inverse of } -11.$$

The numbers $+11$ and -11 are additive inverses (or opposites) of each other. This is true because $(+11) + (-11) = 0$.

FACT FINDING

1. The opposite of a positive number is a _____ number.

2. The opposite of a negative number is a _____ number.

3. The opposite of a number is also called the _____ inverse of the number.

4. The sum of two numbers that are additive inverses of each other is

_____ .

5. If two numbers are additive inverses, they are the same distance from

_____ .

EXERCISES

1. Find the additive inverse of:

    a. -4 **b.** $+100$ **c.** -2 **d.** $\frac{1}{2}$ **e.** $\frac{-1}{4}$ **f.** $-2\frac{3}{8}$

2. Complete each statement so that the result is true.

    a. +6 is the additive inverse of ___.

    b. −5 is the additive inverse of ___.

    c. ___ is the additive inverse of +3.

    d. ___ is the additive inverse of −8.

    e. The additive inverse of +10 is ___.

    f. The opposite of the opposite of 6 is ___.

    g. The opposite of the opposite of −12 is ___.

    h. The opposite of the opposite of the opposite of +9 is ___.

    i. The opposite of the opposite of the opposite of −1 is ___.

3. In each of the following, which number has the greater opposite number?

    a. +7 or +9   **d.** −8 or +5   **g.** +4 or 0   **j.** −15 or −21

    b. −4 or −6   **e.** +10 or −12   **h.** −2 or +2   **k.** −6 or +6

    c. +8 or +5   **f.** 0 or −3   **i.** −14 or −11   **l.** +53 or −62

4. Complete so that each resulting statement will be true.

    a. −21 + 21 = ___   **c.** −60 + ___ = 0   **e.** ___ + (−82) = 0

    b. ___ + (−27) = 0   **d.** 43 + ___ = 0   **f.** ___ + 12 = 0

EXTRA FOR EXPERTS

1. Name the signed number that can replace each question mark and make the resulting statement true.

    a. +23 = +9 + ?   **d** ? + (−30) = +60   **g.** −13 + ? = −4

    b. −17 = −5 + ?   **e.** ? + (+30) = −75   **h.** +25 + ? = 0

    c. +15 = −3 + ?   **f.** +6 + ? = −1   **i.** −41 + ? = +14

2. Explain how the following sets of numbers differ:

    a. The set of negative integers

    b. The set of nonnegative integers

    c. The set of positive integers

    d. The set of nonpositive integers

    e. The set of integers

    f. The set of rational numbers

2-12.1 *Addition of Rational Numbers with Like Signs and Like Denominators*

A number line can be used to find a sum of rational numbers expressed as fractions with the same denominator. For example, by using a number line the sum of $\frac{1}{4}$ and $\frac{2}{4}$ can be found.

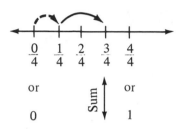

$$\begin{array}{cccc} & \text{or} & \text{Sum} & \text{or} \\ \end{array}$$

Therefore $\frac{1}{4} + \frac{2}{4} = \frac{3}{4}$. 0 1

Similarly, by using a number line the sum of $\frac{-2}{8}$ and $\frac{-3}{8}$ can be found.

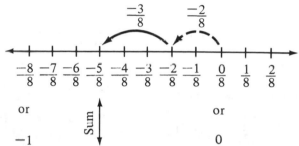

or   Sum   or

Therefore $\frac{-2}{8} + \frac{-3}{8} = \frac{-5}{8}$. -1 0

To find the sum of rational numbers with the same denominator (without using a number line) use the rules below.

Rule	Example
	(a) $\dfrac{1}{4} + \dfrac{2}{4}$
1. Keep the common denominator.	$= \dfrac{}{4}$
2. Find the sum of the numerators, using the rules for adding signed numbers.	$= \dfrac{1 + 2}{4} = \dfrac{3}{4}$
	Answer: $\dfrac{1}{4} + \dfrac{2}{4} = \dfrac{3}{4}$

Rule	Example
	(b) $\dfrac{-2}{8} + \dfrac{-3}{8}$
1. Keep the common denominator.	$= \dfrac{}{8}$
2. Find the sum of the numerators, using the rules for adding signed numbers.	$= \dfrac{-2 + (-3)}{8} = \dfrac{-5}{8}$
	Answer: $\dfrac{-2}{8} + \dfrac{-3}{8} = \dfrac{-5}{8}$

FACT FINDING

To add two rational numbers with the same denominator:

a. Keep the _____ _____ .

b. _____ the numerators, using the rules for _____

_____ _____ .

MODEL PROBLEM

Find each of the following sums:

a. $\left(\dfrac{-22}{13}\right) + \left(\dfrac{-5}{13}\right)$.

Solution: (1) Keep the common denominator: $\dfrac{-22}{13} + \dfrac{-5}{13} = \dfrac{}{13}$.

(2) Find the sum of the numerators, using the rules for adding signed numbers:

$$= \dfrac{(-22) + (-5)}{13}$$

$$= \dfrac{-27}{13}.$$

(3) Answer: $\left(\dfrac{-22}{13}\right) + \left(\dfrac{-5}{13}\right) = \dfrac{-27}{13}$.

b. $\left(-1\dfrac{1}{4}\right) + \left(-2\dfrac{1}{4}\right)$.

Solution: (1) Write each mixed number as an improper fraction:

$$-1\dfrac{1}{4} = \dfrac{-5}{4}$$

$$-2\dfrac{1}{4} = \dfrac{-9}{4}$$

(2) Keep the common denominator and add the numerators, using the rules for adding signed numbers:

$$\dfrac{(-5) + (-9)}{4}$$

$$= \dfrac{-14}{4}.$$

(3) Answer: $\left(-1\dfrac{1}{4}\right) + \left(-2\dfrac{1}{4}\right) = \dfrac{-14}{4}$ or $-3\dfrac{2}{4}$.

EXERCISES

Apply the rules for adding signed numbers and adding fractions with like denominators to find each of the following sums. Express results as proper fractions or mixed numbers.

a. $\dfrac{+20}{10} + \dfrac{+32}{10}$ e. $\dfrac{-7}{9} + \dfrac{-11}{9}$ i. $\left(+23\tfrac{2}{5}\right) + \left(+17\tfrac{3}{5}\right)$

b. $\dfrac{+2}{7} + \dfrac{+6}{7}$ f. $\dfrac{+13}{7} + \dfrac{+10}{7}$ j. $\left(-24\tfrac{3}{8}\right) + \left(-35\tfrac{5}{8}\right)$

c. $\dfrac{-2}{3} + \dfrac{-5}{3}$ g. $\dfrac{-23}{17} + \dfrac{-6}{17}$ k. $\left(+16\tfrac{3}{7}\right) + \left(+9\tfrac{5}{7}\right)$

d. $\dfrac{+3}{7} + \dfrac{+6}{7}$ h. $\dfrac{-11}{16} + \dfrac{-9}{16}$

l. $+18\tfrac{3}{8} + +6\tfrac{5}{8} + +13\tfrac{7}{8}$ m. $-23\tfrac{5}{12} + -12\tfrac{7}{12} + -9\tfrac{11}{12}$

EXTRA FOR EXPERTS

1. If $\dfrac{a}{c}$ and $\dfrac{b}{c}$ represent two rational numbers, represent their sum.

2. Represent the sum of each of the following as a single fraction:

a. $\dfrac{t}{10} + \dfrac{r}{10}$ b. $\dfrac{2}{y} + \dfrac{3}{y}$ c. $\dfrac{9}{2x} + \dfrac{6}{2x}$

2-12.2 Addition of Rational Numbers with Unlike Denominators and Like Signs

STUDY GUIDE To add rational numbers, the *denominators must be the same*. In order to change a fraction to an equivalent fraction, multiply the numerator and the denominator of the given fraction by the same nonzero number.

Example: $\dfrac{1}{2} = \dfrac{1 \cdot 3}{2 \cdot 3} = \dfrac{3}{6}$.

The least common multiple of the denominators (or least common denominator) must also be found. The fractions are then written as equivalent fractions, each with the least common multiple as its denominator. To add the equivalent fractions with like denominators, use this rule: (1) Keep the common denominator, and (2) add the numerators, using the rules for adding signed numbers.

1. To add rational numbers the denominators must be _____

_____ .

2. The least common multiple of the denominators becomes the

_____ _____ denominator.

3. To add rational numbers with the same denominator:

a. _____ _____ _____ _____

b. _____ _____ _____ _____

_____ _____ _____ _____

_____ _____ .

MODEL PROBLEMS

1. Find the sum of $\dfrac{-1}{2}$ and $\dfrac{-3}{8}$.

Solution: (1) Find the multiples of each denominator:

Multiples of 2:
2, 4, 6, ⑧, 10, ...
Multiples of 8: ⑧, 16, 24, ...

(2) Find the LCM (least common multiple):

LCM is 8.

(3) Change each fraction into an equivalent fraction with the LCM as its denominator:

$$\dfrac{-1}{2} = \dfrac{-1 \cdot 4}{2 \cdot 4} = \dfrac{-4}{8}$$
$$+\dfrac{-3}{8} = \dfrac{-3 \cdot 1}{8 \cdot 1} = \dfrac{-3}{8}$$

(4) Add the numerators, using the rules for adding signed numbers. Keep the common denominator.

$$\text{Sum} = \dfrac{-7}{8}.$$

(5) Answer: $\dfrac{-1}{2} + \dfrac{-3}{8} = \dfrac{-7}{8}.$

2. Find the sum of $-3\dfrac{4}{5} + -1\dfrac{1}{2}$.

Solution: (1) Write each mixed number as an improper fraction:

$$-3\dfrac{4}{5} = \dfrac{-19}{5}$$
$$-1\dfrac{1}{2} = \dfrac{-3}{2}$$

(2)	Find the multiples of each denominator:	Multiples of 5: 5, ⑩, 15, 20, . . . Multiples of 2: 2, 4, 6, 8, ⑩, 12, . . .
(3)	Find the LCM:	LCM = 10.
(4)	Change each fraction into an equivalent fraction with the LCM as its denominator:	$\dfrac{-19}{5} = \dfrac{-19 \cdot 2}{5 \cdot 2} = \dfrac{-38}{10}$ $\dfrac{-3}{2} = \dfrac{-3 \cdot 5}{2 \cdot 5} = \dfrac{-15}{10}$
(5)	Keep the common denominator and add the numerators, using the rules for adding signed numbers:	$\dfrac{(-38) + (-15)}{10} = \dfrac{-53}{10}$
(6)	Answer:	$-3\dfrac{4}{5} + -1\dfrac{1}{2} = \dfrac{-53}{10}$ or $-5\dfrac{3}{10}$.

EXERCISES

1. Find the sum for each of the following:

a. $\left(\dfrac{+2}{4}\right) + \left(\dfrac{+2}{3}\right)$ **h.** $\left(\dfrac{5}{6}\right) + \left(\dfrac{1}{8}\right)$ **o.** $\left(-8\dfrac{1}{7}\right) + \left(-9\dfrac{5}{6}\right)$

b. $\left(\dfrac{-1}{5}\right) + \left(\dfrac{-1}{3}\right)$ **i.** $\left(\dfrac{-1}{2}\right) + \left(\dfrac{-2}{3}\right)$ **p.** $\left(-6\dfrac{2}{5}\right) + \left(-3\dfrac{1}{2}\right)$

c. $\left(\dfrac{-7}{12}\right) + \left(\dfrac{-5}{6}\right)$ **j.** $\left(\dfrac{+1}{4}\right) + \left(\dfrac{+3}{10}\right)$ **q.** $\left(-4\dfrac{1}{5}\right) + \left(-5\dfrac{1}{4}\right)$

d. $\left(\dfrac{3}{10}\right) + \left(\dfrac{2}{5}\right)$ **k.** $\left(\dfrac{+7}{2}\right) + \left(\dfrac{+4}{5}\right)$ **r.** $\left(-37\dfrac{3}{7}\right) + \left(-29\dfrac{1}{4}\right)$

e. $\left(\dfrac{-3}{4}\right) + \left(\dfrac{-3}{5}\right)$ **l.** $\left(\dfrac{+4}{9}\right) + \left(\dfrac{1}{6}\right)$ **s.** $\left(-45\dfrac{3}{7}\right) + \left(-29\dfrac{1}{4}\right)$

f. $\left(\dfrac{-4}{9}\right) + \left(\dfrac{-1}{3}\right)$ **m.** $\left(+3\dfrac{1}{7}\right) + \left(\dfrac{+1}{8}\right)$

g. $\left(\dfrac{-2}{5}\right) + \left(\dfrac{-3}{8}\right)$ **n.** $\left(\dfrac{+4}{7}\right) + \left(+9\dfrac{5}{9}\right)$

2. Find the sum for each of the following:

a. $\left(\dfrac{+3}{5}\right) + \left(\dfrac{+1}{2}\right) + \left(\dfrac{+7}{10}\right)$ **d.** $\left(-5\dfrac{1}{4}\right) + \left(-4\dfrac{2}{3}\right) + \left(-2\dfrac{5}{8}\right)$

b. $\left(\dfrac{-1}{4}\right) + \left(\dfrac{-11}{12}\right) + \left(\dfrac{-9}{16}\right)$ **e.** $\left(-4\dfrac{1}{6}\right) + \left(-5\dfrac{1}{2}\right) + \left(-10\dfrac{7}{8}\right)$

c. $\left(+3\dfrac{1}{2}\right) + \left(+7\dfrac{3}{4}\right) + \left(+4\dfrac{1}{3}\right)$ **f.** $\left(+6\dfrac{1}{4}\right) + \left(+4\dfrac{7}{12}\right) + \left(+2\dfrac{3}{16}\right)$

3. Write the rules for adding:

    a. Fractions with like denominators

    b. Fractions with unlike denominators

    c. Signed numbers with like signs

4. Solve each of the following problems:

    a. A recipe calls for $\frac{1}{2}$ cup of granulated white sugar and $\frac{3}{4}$ cup of brown sugar. How much sugar will be used for this recipe?

    b. A postman measured the number of miles he walked in three days. He covered $\frac{7}{10}$ mile, $\frac{1}{2}$ mile, and $\frac{9}{15}$ mile. Find the total number of miles he walked.

EXTRA FOR EXPERTS

1. If $p = \frac{+11}{12}$, $q = \frac{7}{8}$, and $r = \frac{1}{4}$, evaluate:

    a. $p + q + r$ **c.** $-q + p + r$

    b. $-r + p + q$ **d.** $-p + r + q$

Find the answers to the following problems by means of signed numbers:

2. A certain stock closed on Monday at a selling price of $37\frac{3}{4}$. Find the closing price of the stock on Friday if it gained $\frac{7}{8}$ point on Tuesday, lost $1\frac{1}{4}$ points on Wednesday, lost $2\frac{1}{2}$ points on Thursday, and gained $\frac{3}{8}$ point on Friday.

3. A typist using $8\frac{1}{2}$-inch by 11-inch paper wants to leave a top margin of $1\frac{1}{2}$ inches, a bottom margin of $\frac{3}{4}$ inch, and two side margins of $\frac{7}{8}$ inch each. What are the dimensions of the typed region?

4. Find the missing numbers.

    a. $? + \left(\frac{+2}{3}\right) = +1\frac{1}{6}$ **b.** $\left(+2\frac{2}{3}\right) + ? = +7\frac{5}{6}$ **c.** $\left(-3\frac{3}{8}\right) + ? = -9\frac{3}{4}$

2-13.1 *Addition of Rational Numbers with Like Denominators and Unlike Signs*

The number line can be used to help develop rules to add rational numbers with like denominators and unlike signs.

Example: Find the sum of $\dfrac{-7}{8}$ and $\dfrac{+3}{8}$.

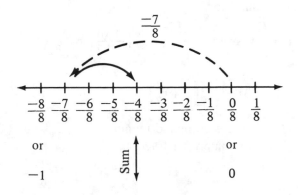

Therefore $\dfrac{-7}{8} + \dfrac{+3}{8} = \dfrac{-4}{8}$ or $\dfrac{-1}{2}$.

This demonstrates that the rule for adding rational numbers with like denominators and unlike signs is the same as the rule for adding rational numbers with unlike denominators and like signs, that is: (1) Keep the common denominator, and (2) add the numerators, using the rules for adding signed numbers.

The addition of rational numbers with unlike signs is equivalent to the subtraction of fractions studied in arithmetic.

Example: $\dfrac{9}{7} - \dfrac{2}{7} = \dfrac{5}{7}$ is the same as $\left(\dfrac{9}{7}\right) + \left(\dfrac{-2}{7}\right) = \dfrac{5}{7}$.

FACT FINDING

1. To add two rational numbers with like denominators:

    a. _____ the common denominator.

    b. _____ the numerators, using the rules for _____

    _____ _____ .

2. Subtraction of arithmetic fractions is equivalent to _____

of rational numbers with _____ signs.

1. Find the sum: $\left(\dfrac{-27}{32}\right) + \left(\dfrac{+18}{32}\right)$.

   Solution: (1) Keep the common denominator:

   $$\overline{32}$$

   (2) Add the numerators, using the rules for adding signed numbers:

   $$\dfrac{(-27) + (+18)}{32} = \dfrac{-9}{32}$$

   (3) Answer: $\left(\dfrac{-27}{32}\right) + \left(\dfrac{+18}{32}\right) = \dfrac{-9}{32}$.

2. Find the sum: $\left(+31\dfrac{7}{8}\right) + \left(-19\dfrac{2}{8}\right)$.

   Solution: (1) Write each mixed number as an improper fraction:

   $$+31\dfrac{7}{8} = \dfrac{+255}{8}$$
   $$-19\dfrac{2}{8} = \dfrac{-154}{8}$$

   (2) Keep the common denominator:

   $$\overline{8}$$

   (3) Add the numerators, using the rules for adding signed numbers:

   $$\dfrac{(+255) + (-154)}{8}$$

   (4) Answer: $\left(+31\dfrac{7}{8}\right) + \left(-19\dfrac{2}{8}\right) = \dfrac{+101}{8}$ or $+12\dfrac{5}{8}$.

3. Subtract: $\dfrac{9}{16} - \dfrac{3}{16}$ (using the rules studied in arithmetic).

   Solution: (1) Keep the common denominator:

   $$\overline{16}$$

   (2) Find the difference between the numerators:

   $$\dfrac{9 - 3}{16} = \dfrac{6}{16}$$

   (3) Answer: $\dfrac{9}{16} - \dfrac{3}{16} = \dfrac{6}{16}$.

1. Find the sum for each of the following:

   a. $\left(\dfrac{-1}{4}\right) + \left(\dfrac{+11}{4}\right)$ f. $\left(\dfrac{-7}{20}\right) + \left(\dfrac{5}{20}\right)$ k. $\left(5\dfrac{1}{3}\right) + \left(-6\dfrac{2}{3}\right)$

   b. $\left(\dfrac{+5}{12}\right) + \left(\dfrac{-3}{12}\right)$ g. $\left(\dfrac{+1}{4}\right) + \left(\dfrac{-11}{4}\right)$ l. $\left(-1\dfrac{1}{7}\right) + \left(+3\dfrac{2}{7}\right)$

   c. $\left(\dfrac{-3}{4}\right) + \left(\dfrac{1}{4}\right)$ h. $\left(+40\dfrac{5}{6}\right) + \left(-20\dfrac{1}{6}\right)$ m. $\left(-2\dfrac{3}{5}\right) + \left(+1\dfrac{1}{5}\right)$

   d. $\left(\dfrac{6}{9}\right) + \left(\dfrac{-3}{9}\right)$ i. $\left(-18\dfrac{9}{10}\right) + \left(+8\dfrac{3}{10}\right)$ n. $(+4) + \left(-9\dfrac{5}{8}\right)$

   e. $\left(\dfrac{2}{12}\right) + \left(\dfrac{-8}{12}\right)$ j. $\left(-2\dfrac{1}{2}\right) + \left(+3\dfrac{1}{2}\right)$ o. $(+2) + \left(-6\dfrac{1}{2}\right)$

2. Subtract the following arithmetic fractions:

   a. $\dfrac{7}{10} - \dfrac{3}{10}$ c. $\dfrac{21}{32} - \dfrac{15}{32}$ e. $\dfrac{103}{100} - \dfrac{47}{100}$

   b. $\dfrac{5}{6} - \dfrac{1}{6}$ d. $\dfrac{53}{64} - \dfrac{29}{64}$ f. $\dfrac{214}{200} - \dfrac{132}{200}$

3. Solve each of the following problems:

   a. Mrs. Nilson found that she weighed $124\dfrac{3}{4}$ pounds. Later she lost $11\dfrac{1}{4}$ pounds. What is her new weight?

   b. Mr. Gray had only $8\dfrac{7}{8}$ hours for his business trip. He used $4\dfrac{3}{8}$ hours for traveling time. How much time did he have left for his business activities?

EXTRA FOR EXPERTS

Find the value of n in each statement that will make the statement true.

a. $\dfrac{5}{4} + \dfrac{n}{4} = \dfrac{15}{4} + \dfrac{3}{4}$ c. $\dfrac{n}{12} - \dfrac{17}{12} = \dfrac{19}{12} - \dfrac{11}{12}$

b. $\dfrac{18}{7} - \dfrac{n}{7} = \dfrac{5}{7} + \dfrac{3}{7}$ d. $n + 4\dfrac{7}{8} = 7\dfrac{5}{8}$

2-13.2 *Addition of Rational Numbers with Unlike Denominators*

STUDY GUIDE To add two rational numbers with unlike denominators, first find the LCD (or LCM) and then convert each rational number into an equivalent rational number with the LCD as the new denominator. Also use the rule developed previously to add rational numbers with like denominators. See the Model Problems below.

MODEL PROBLEMS

1. Find the sum: $\left(\dfrac{+3}{2}\right) + \left(\dfrac{-4}{5}\right)$.

    Solution: (1) Find the multiples of each denominator:

    Multiples of 2:
    2, 4, 6, 8, ⑩, ...
    Multiples of 5:
    5, ⑩, 15, ...

    (2) Find the LCM (or LCD): LCD (or LCM) = 10.

    (3) Convert each fraction into an equivalent fraction having the LCD as its new denominator:

    $$\frac{+3}{2} = \frac{+3 \cdot 5}{2 \cdot 5} = \frac{+15}{10}$$

    $$\frac{-4}{5} = \frac{-4 \cdot 2}{5 \cdot 2} = \frac{-8}{10}$$

    (4) Keep the common denominator and add the numerators, using the rules for adding signed numbers:

    $$= \frac{(+15) + (-8)}{10}$$

    $$= \frac{+7}{10}$$

    (5) Answer: $\left(\dfrac{+3}{2}\right) + \left(\dfrac{-4}{5}\right) = \dfrac{+7}{10}$.

2. Find the sum: $\left(-8\dfrac{9}{10}\right) + \left(+6\dfrac{5}{6}\right)$.

    Solution: (1) Write each mixed number as an improper fraction:

    $$-8\frac{9}{10} = \frac{-89}{10}$$

    $$+6\frac{5}{6} = \frac{+41}{6}$$

    (2) Find the multiples of each denominator:

    Multiples of 10:
    10, 20, ㉚, 40, 50, . . .
    Multiples of 6:
    6, 12, 18, 24, ㉚, . . .

    (3) Find the LCM (or LCD): LCM (or LCD) = 30.

(4) Convert each fraction into an equivalent fraction having the LCD as its new denominator:	$\dfrac{-89}{10} = \dfrac{-89 \cdot 3}{10 \cdot 3} = \dfrac{-267}{30}$ $\dfrac{+41}{6} = \dfrac{+41 \cdot 5}{6 \cdot 5} = \dfrac{+205}{30}$
(5) Keep the common denominator and add the numerators, using the rules for adding signed numbers:	$= \dfrac{(-267) + (205)}{30}$ $= \dfrac{-62}{30}$
(6) Answer:	$\left(-8\dfrac{9}{10}\right) + \left(+6\dfrac{5}{6}\right) = \dfrac{-62}{30}$ or $-2\dfrac{2}{30}$.

EXERCISES

1. Find the sum for each of the following:

a. $\left(\dfrac{-5}{6}\right) + \left(\dfrac{+2}{3}\right)$ 　　**g.** $\left(\dfrac{-3}{7}\right) + \left(\dfrac{1}{3}\right)$ 　　**m.** $\left(-7\dfrac{2}{7}\right) + \left(+5\dfrac{1}{4}\right)$

b. $\left(\dfrac{3}{4}\right) + \left(\dfrac{-1}{2}\right)$ 　　**h.** $\left(\dfrac{+7}{9}\right) + \left(\dfrac{-1}{3}\right)$ 　　**n.** $\left(-2\dfrac{7}{8}\right) + \left(+1\dfrac{7}{12}\right)$

c. $\left(\dfrac{5}{6}\right) + \left(\dfrac{-7}{8}\right)$ 　　**i.** $\left(\dfrac{+1}{5}\right) + \left(\dfrac{-1}{3}\right)$ 　　**o.** $\left(-14\dfrac{1}{3}\right) + \left(+7\dfrac{1}{6}\right)$

d. $\left(\dfrac{-3}{8}\right) + \left(\dfrac{7}{10}\right)$ 　　**j.** $\left(\dfrac{-1}{4}\right) + \left(\dfrac{+5}{6}\right)$ 　　**p.** $\left(+32\dfrac{2}{3}\right) + \left(-31\dfrac{1}{6}\right)$

e. $\left(\dfrac{-5}{6}\right) + \left(\dfrac{9}{15}\right)$ 　　**k.** $\left(\dfrac{1}{18}\right) + \left(\dfrac{-1}{12}\right)$ 　　**q.** $\dfrac{3}{8} + \dfrac{1}{4} + \left(\dfrac{-5}{16}\right)$

f. $\left(\dfrac{2}{3}\right) + \left(\dfrac{-1}{2}\right)$ 　　**l.** $\left(\dfrac{-3}{10}\right) + \left(\dfrac{+6}{25}\right)$ 　　**r.** $\dfrac{-3}{4} + \dfrac{5}{3} + \left(\dfrac{-2}{12}\right)$

2. Subtract the following arithmetic fractions:

a. $\dfrac{2}{3} - \dfrac{1}{4}$ 　　**c.** $\dfrac{9}{10} - \dfrac{2}{5}$ 　　**e.** $\dfrac{5}{6} - \dfrac{3}{8}$ 　　**g.** $\dfrac{7}{12} - \dfrac{3}{16}$

b. $\dfrac{7}{8} - \dfrac{1}{6}$ 　　**d.** $\dfrac{1}{4} - \dfrac{1}{5}$ 　　**f.** $\dfrac{5}{8} - \dfrac{1}{12}$ 　　**h.** $\dfrac{11}{16} - \dfrac{9}{20}$

EXTRA FOR EXPERTS

Find the result of each of the following:

a. $\left(\dfrac{7}{8} + \dfrac{11}{12}\right) - \dfrac{2}{3}$ 　　**b.** $\left(\dfrac{1}{6} + \dfrac{3}{8}\right) - \dfrac{1}{4}$ 　　**c.** $\dfrac{17}{12} - \left(\dfrac{1}{6} + \dfrac{5}{18}\right)$

2-13.3 *Subtraction of Rational Numbers Involving Regrouping*

Regrouping or exchanging is necessary in subtracting a mixed number from a whole number. Regrouping or exchanging is sometimes necessary in subtracting a mixed number from a mixed number. Subtraction of these arithmetic numbers is equivalent to the addition of rational numbers with unlike signs.

Examples: $8 - \dfrac{2}{5}$ is the same as $+8 + \dfrac{-2}{5}$.

$7\dfrac{3}{5} - 1\dfrac{4}{5}$ is the same as $+7\dfrac{3}{5} + \left(-1\dfrac{4}{5}\right)$.

To avoid regrouping, it is necessary to use the fact that subtraction of these arithmetic numbers is equivalent to the addition of rational numbers with unlike signs. This is illustrated in each of the model problems given below.

MODEL PROBLEM

Perform the following subtractions of arithmetic numbers.

a. $8 - \dfrac{2}{5}$.

Solution:

(1) Write 8 as an equivalent fraction having a denominator of 5: $\qquad \dfrac{8}{1} = \dfrac{8 \cdot 5}{1 \cdot 5} = \dfrac{40}{5}$

(2) Subtract. Subtraction of arithmetic numbers is equivalent to the addition of rational numbers with unlike signs:

$$8 - \dfrac{2}{5} = 8 + \left(\dfrac{-2}{5}\right)$$
$$= \dfrac{40}{5} + \left(\dfrac{-2}{5}\right)$$

(3) Keep the common denominator and add the numerators, using the rules for adding signed numbers:

$$= \dfrac{38}{5}$$

(4) Answer: $\quad 8 - \dfrac{2}{5} = \dfrac{38}{5}$ or $7\dfrac{3}{5}$.

b. $6\dfrac{2}{7} - 4\dfrac{4}{7}$.

Solution:

(1) Write each mixed number as an improper fraction:

$$6\dfrac{2}{7} = \dfrac{44}{7}$$
$$4\dfrac{4}{7} = \dfrac{32}{7}$$

(2) Use the fact that subtraction of arithmetic numbers is equivalent to the addition of rational numbers with unlike signs:

$$\dfrac{44}{7} - \dfrac{32}{7} = \dfrac{44}{7} + \left(\dfrac{-32}{7}\right)$$

(3) Keep the common denominator and add the numerators, using the rules for adding signed numbers:

$$= \frac{12}{7}$$

(4) Answer: $6\frac{2}{7} - 4\frac{4}{7} = \frac{12}{7}$ or $1\frac{5}{7}$.

c. $5\frac{3}{4} - 1\frac{7}{8}$.

Solution:

(1) Write each mixed number as an improper fraction:

$$5\frac{3}{4} = \frac{23}{4}$$

$$1\frac{7}{8} = \frac{15}{8}$$

(2) Use the fact that subtraction of arithmetic numbers is equivalent to the addition of rational numbers with unlike signs:

$$\frac{23}{4} - \frac{15}{8} = \frac{23}{4} + \left(\frac{-15}{8}\right)$$

(3) Find the multiples of each denominator:

Multiples of 4:
    4, ⑧, 12, 16, . . .
Multiples of 8:
    ⑧, 16, 24, . . .

(4) Find the LCM:

LCM = 8.

(5) Convert each fraction into an equivalent fraction having the LCM (or LCD) as its new denominator:

$$\frac{23}{4} = \frac{23 \cdot 2}{4 \cdot 2} = \frac{46}{8}$$

(6) Keep the common denominator and add the numerators, using the rules for adding signed numbers:

$$\frac{46}{8} + \frac{-15}{8} = \frac{31}{8}$$

(7) Answer: $5\frac{3}{4} - 1\frac{7}{8} = \frac{31}{8}$ or $3\frac{7}{8}$.

EXERCISES

1. Supply the missing numeral so that each resulting statement will be true.

    a. $9\frac{1}{10} = 8\frac{}{10}$ **c.** $11\frac{3}{6} = 10\frac{}{6}$ **e.** $6 = 5\frac{}{4}$

    b. $5\frac{3}{8} = 4\frac{}{8}$ **d.** $14\frac{7}{25} = 13\frac{}{25}$ **f.** $6\frac{7}{11} = 5\frac{}{11}$

2. Do the following subtraction problems:

   a. $5 - \dfrac{3}{8}$ **h.** $11 - 5\dfrac{4}{9}$ **o.** $9\dfrac{3}{7} - 1\dfrac{5}{7}$ **v.** $10\dfrac{1}{2} - 4\dfrac{5}{6}$

   b. $9 - \dfrac{7}{12}$ **i.** $9 - 2\dfrac{1}{4}$ **p.** $16\dfrac{9}{16} - 7\dfrac{12}{16}$ **w.** $12\dfrac{1}{4} - 6\dfrac{1}{2}$

   c. $12 - \dfrac{9}{10}$ **j.** $42 - 17\dfrac{7}{20}$ **q.** $11\dfrac{3}{10} - 6\dfrac{9}{10}$ **x.** $7\dfrac{1}{3} - 4\dfrac{1}{2}$

   d. $10 - \dfrac{2}{3}$ **k.** $27 - 20\dfrac{13}{16}$ **r.** $36\dfrac{1}{5} - 18\dfrac{4}{5}$ **y.** $13\dfrac{1}{4} - 7\dfrac{2}{3}$

   e. $5 - 1\dfrac{3}{4}$ **l.** $37 - 13\dfrac{3}{14}$ **s.** $8\dfrac{11}{20} - 3\dfrac{14}{20}$ **z.** $9\dfrac{3}{4} - 1\dfrac{9}{10}$

   f. $7 - 4\dfrac{2}{3}$ **m.** $5\dfrac{1}{6} - 2\dfrac{5}{6}$ **t.** $17\dfrac{5}{32} - 13\dfrac{29}{32}$

   g. $17 - 12\dfrac{1}{3}$ **n.** $7\dfrac{3}{8} - 4\dfrac{5}{8}$ **u.** $9\dfrac{3}{8} - 4\dfrac{3}{4}$

3. Add the following rational numbers:

   a. $\left(+20\dfrac{1}{4}\right) + \left(-10\dfrac{3}{8}\right)$ **c.** $\left(-6\dfrac{3}{8}\right) + \left(+1\dfrac{3}{4}\right)$ **e.** $\left(-12\dfrac{3}{4}\right) + \left(+9\dfrac{4}{5}\right)$

   b. $\left(+4\dfrac{1}{8}\right) + \left(\dfrac{-3}{16}\right)$ **d.** $\left(-15\dfrac{2}{7}\right) + \left(+20\dfrac{1}{4}\right)$ **f.** $\left(-8\dfrac{1}{6}\right) + \left(+4\dfrac{3}{8}\right)$

4. Solve each of the following problems:

   a. Mary is supposed to practice 12 hours this weekend for a tennis match. She practiced $3\dfrac{1}{4}$ hours on Saturday afternoon, $2\dfrac{1}{2}$ hours that evening, and $1\dfrac{3}{4}$ hours on Sunday morning. How many more hours must she practice on Sunday?

   b. A carpenter had a piece of lumber 18 ft. long. Pieces that measured $6\dfrac{1}{4}$ ft., $7\dfrac{3}{4}$ ft., $\dfrac{1}{2}$ ft., and 2 ft. were cut from this board. How much was used? How much of the original piece of lumber did the carpenter have left?

   c. Alice bought $4\dfrac{3}{8}$ yards of material. If she plans to use $3\dfrac{3}{4}$ yards for a suit, how much extra material did Alice buy?

   d. It took Mrs. Colmer 5 hours to make a round trip to Middletown. The trip out took $2\dfrac{3}{4}$ hours. How many hours did the return trip take?

   e. Mr. Chase took $9\dfrac{1}{2}$ hours to paint the kitchen and the living room in his house. It took him $2\dfrac{3}{4}$ hours to paint the kitchen alone. How many hours did it take him to paint the living room?

1. Find the value of each of the following numerical expressions:

   a. $(6 + 9) + (8 - 8) + (3 - 10)$ c. $(2 - 5) - (11 - 3) + (1 - 8)$

   b. $(7 - 4) + (4 - 7) + (6 - 14)$ d. $10 + (5 - 7) - 6 + (2 - 9)$

2. If a certain number is subtracted from $\frac{2}{3}$, the result is $\frac{-1}{4}$. What is the number?

3. Compute the sum for each of the following:

   a. $\frac{1}{2} + \frac{4}{5} + \left(\frac{-3}{2}\right) + \left(\frac{-9}{5}\right)$

   g. $\left(-6\frac{3}{4}\right) + (-5) + \left(4\frac{3}{4}\right)$

   b. $\frac{-2}{15} + \frac{5}{6} + \left(\frac{-1}{6}\right) + \frac{7}{15}$

   h. $\left(7\frac{3}{4}\right) + \left(-4\frac{1}{4}\right) + \left(-6\frac{3}{4}\right)$

   c. $\frac{1}{4} + \left(\frac{-7}{10}\right) + \frac{3}{4} + \frac{1}{5}$

   i. $-1\frac{1}{6} + 2\frac{7}{8} + \left(\frac{-3}{4}\right)$

   d. $\frac{2}{3} + \left(\frac{-5}{4}\right) + \left(\frac{-1}{6}\right) + \frac{3}{4}$

   j. $6\frac{1}{4} + \left(-4\frac{7}{12} + 2\frac{3}{16}\right)$

   e. $\left(-2\frac{1}{2}\right) + \left(3\frac{1}{2}\right) + \left(-8\frac{1}{2}\right)$

   k. $4\frac{2}{3} + 1\frac{1}{2} + \left(-3\frac{1}{4}\right)$

   f. $\left(-3\frac{1}{2}\right) + 5\frac{1}{2} + \left(-9\frac{1}{2}\right)$

   l. $-4\frac{1}{6} + \left(15\frac{1}{2} - 10\frac{7}{8}\right)$

Measuring Your Progress

SECTION 2-1

1. Which is an inequality?

   a. $-3y > -21$ b. $9y - 3 = -21$

2. Write symbolically:

   a. Twenty-five less eleven is less than fifteen.

   b. Twice eighteen is greater than forty divided by two.

   c. x increased by five is equal to thirty.

   d. y decreased by two is less than twelve.

3. Write in words:

    a. $6 - 2 < 17$ **b.** $75 \div 5 < 8 \cdot 4$ **c.** $6x + 7 > 43$

4. Is each of the following TRUE or FALSE?

    a. $4(8) - 3 > 5^2$ **b.** $6 \cdot 9 = 54$

5. Insert the symbol "<," ">," or "=" to make each statement true.

    a. $60 + 37 \_\_ 90$ **b.** $41 - 17 \_\_ 3(9)$ **c.** $(18 - 3) + 7 \_\_ 7.5 + 15 \div 3$

SECTION 2-2

1. What number is:

    a. 3 units to the right of 8?   **b.** 4 units to the left of 9?

2. For each pair of numbers name the number that will appear farther to the right on the number line, and use an inequality symbol to show the relationship between each pair of numbers.

    a. 7, 3   **b.** 0, 8   **c.** $\dfrac{1}{4}, \dfrac{3}{4}$ **d.** $2, \dfrac{4}{4}$

3. Which of the following is a mixed number?

    a. $2\dfrac{2}{3}$ **b.** 7   **c.** $\dfrac{4}{7}$ **d.** 0

4. Which of the following is a proper fraction?

    a. $\dfrac{7}{5}$ **b.** $\dfrac{16}{16}$ **c.** $\dfrac{7}{8}$ **d.** 0

SECTION 2-3

1. Which number is at point P on the number line below?

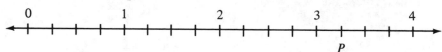

    a. $1\dfrac{3}{4}$ **b.** $2\dfrac{1}{4}$ **c.** $3\dfrac{1}{4}$ **d.** $3\dfrac{1}{2}$

2. Which letter shows $2\dfrac{1}{4}$ inches on the ruler below?

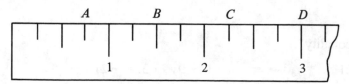

    a. A **b.** B **c.** C **d.** D

3. **a.** How many $\dfrac{1}{8}$ inches are there in $1\dfrac{1}{8}$ inches?

    b. How many $\dfrac{1}{8}$ inches are there in 4 inches?

c. How many $\frac{1}{4}$ inches are there in $4\frac{1}{2}$ inches?

d. How many $\frac{1}{2}$ inches are there in $5\frac{1}{2}$ inches?

e. How many $\frac{1}{2}$ inches are there in 3 inches?

4. Multiplying or dividing the numerator and denominator of a fraction by the same number, except zero, does not change the value of the fraction. TRUE or FALSE?

5. In order to rename $\frac{2}{3}$ as $\frac{12}{18}$ you:

    a. add 10   **b.** multiply by 6   **c.** multiply by $\frac{6}{6}$ **d.** add 15

6. In each of the following, supply the missing numeral so that the fractions are equivalent:

    a. $\frac{5}{8} = \frac{}{48}$ **b.** $\frac{4}{9} = \frac{}{81}$ **c.** $\frac{2}{2} = \frac{}{8}$ **d.** $\frac{}{4} = \frac{12}{16}$

7. Which of the following is an improper fraction?

    a. $\frac{1}{2}$ **b.** $\frac{21}{7}$ **c.** $\frac{8}{5}$ **d.** $\frac{36}{9}$

8. Change each of the following to an improper fraction:

    a. $2\frac{3}{4}$ **b.** $5\frac{2}{3}$ **c.** $6\frac{7}{10}$ **d.** $2\frac{13}{16}$

9. Change each of the following to a mixed number or a whole number:

    a. $\frac{17}{6}$ **b.** $\frac{35}{8}$ **c.** $\frac{42}{7}$ **d.** $\frac{15}{3}$

10. Insert the symbol "<" or ">" between the two numerals so that each resulting statement will be true.

    a. $\frac{45}{6} - 7\frac{1}{6}$ **c.** $\frac{7}{11} - \frac{3}{11}$ **e.** $\frac{8}{8} - \frac{3}{4}$

    b. $\frac{25}{4} - 6\frac{3}{4}$ **d.** $\frac{1}{8} - \frac{1}{2}$ **f.** $\frac{7}{16} - \frac{5}{8}$

11. Arrange in order of value from smaller to larger:

    a. $\frac{4}{9}, \frac{2}{9}, \frac{8}{9}, \frac{5}{9}$ **b.** $\frac{9}{11}, \frac{0}{11}, \frac{3}{11}, \frac{11}{11}$

12. Which of the fractions $\frac{3}{2}, \frac{10}{8}$, and $\frac{7}{4}$ represents the greatest value?

SECTION 2-4

1. Choose the correct value of the points shown on the ruler below:

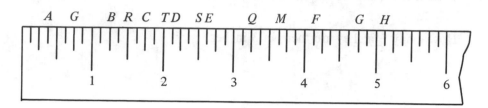

a. Point A   (1) $\frac{3}{16}''$ (2) $\frac{3}{8}''$ (3) $\frac{1}{2}''$ (4) $\frac{3}{12}''$ (5) none

b. Point B   (1) $1\frac{1}{8}''$ (2) $1\frac{1}{4}''$ (3) $1\frac{1}{3}''$ (4) $1\frac{2}{3}''$ (5) none

c. Point C   (1) $1\frac{1}{4}''$ (2) $1\frac{3}{8}''$ (3) $1\frac{1}{2}''$ (4) $1\frac{3}{4}''$ (5) none

d. Point D   (1) $2\frac{3}{4}''$ (2) $2\frac{1}{3}''$ (3) $2\frac{1}{4}''$ (4) $2\frac{1}{8}''$ (5) none

e. Point E   (1) $2\frac{1}{2}''$ (2) $2\frac{1}{3}''$ (3) $2\frac{5}{8}''$ (4) $2\frac{6}{8}''$ (5) none

f. Point F   (1) $4\frac{1}{8}''$ (2) $4\frac{1}{4}''$ (3) $4\frac{3}{4}''$ (4) $4\frac{1}{16}''$ (5) none

g. Point G   (1) $4\frac{3}{8}''$ (2) $4\frac{5}{8}''$ (3) $4\frac{3}{16}''$ (4) $4\frac{1}{4}''$ (5) none

h. Point H   (1) $5\frac{3}{4}''$ (2) $5\frac{1}{8}''$ (3) $5\frac{3}{8}''$ (4) $5\frac{3}{16}''$ (5) none

2. Using the ruler drawn above, find the distance from:
   a. *G* to *T*   b. *R* to *S*   c. *T* to *M*   d. *R* to *Q*   e. *G* to *S*

3. Which of the following is closest to 3 inches?
   a. $2\frac{7}{8}''$ b. $2\frac{15}{16}''$ c. $2\frac{3}{4}''$ d. $3\frac{1}{4}''$

4. Which of the following is closest to 4 inches?
   a. $3\frac{7}{8}''$ b. $4\frac{1}{4}''$ c. $3\frac{15}{16}''$ d. $3\frac{3}{4}''$

5. TRUE or FALSE? A straight line has definite length.

6. Complete correctly: Only one straight line can be drawn through two
   _____ .

7. A line segment has: (1) no endpoints (2) one endpoint (3) two endpoints.

8. Referring to the diagram below, name all the line segments in the figure.

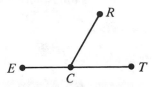

9. Referring to the diagram below:

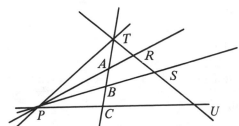

a. Name a point that is on 4 of the lines.

b. Name a point that is on exactly 3 of the lines.

c. How many points are between T and C? Name 2 of them.

d. Name 4 line segments on TU.

10. Measure the line segments below to the nearest $\frac{1''}{8}$ of an inch.

a. └──────────────────┘   **b.** └────┘

c. └────────────────────┘

11. Mark with a "dot" the midpoint of AB. A └──────────────────┘ B

12. Replace the question mark by the numeral that makes each statement true.

a. $\frac{1}{4} = \frac{?}{8}$ **c.** $\frac{1}{2} = \frac{?}{4}$ **e.** $2\frac{1}{2} = \frac{?}{8}$

b. $3 = \frac{?}{4}$ **d.** $\frac{3}{4} = \frac{?}{8}$ **f.** $3\frac{1}{2} = \frac{?}{4}$

13. Use your ruler to measure each side of the polygons below to the nearest $\frac{1''}{8}$ inch. Name the sides of each figure from smallest to largest.

a.   **b.**

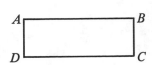   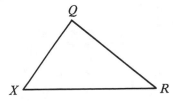

SECTION 2-5

1. State the word that is opposite in meaning to each of the following:

a. Gain   **d.** Slower   **g.** Clockwise

b. Above   **e.** Longer   **h.** Credit

c. Earnings   **f.** Younger   **i.** Asset

2. State the quantity represented by:

    a. A withdrawal of $15   **d.** No change in temperature

    b. A gain of $36   **e.** 8 miles above sea level

    c. Earnings of $450

3. If $+8°$ means 8 degrees above zero, what does $-7°$ mean?

4. If -3% means a decrease of 3% in the cost of living, what does $+9\%$ represent?

5. If -12 means 12 yards lost, what does $+8$ mean?

6. If $+13$ means 13 steps forward, what does -5 mean?

7. Which of the following is represented by negative 8?

    a. 8 miles south   **c.** A gain of $8

    b. 8 miles east   **d.** 8 miles above sea level

8. What will be the result if you:

    a. gain 4 kg and lose 5 kg?   **d.** win $25 and find $10

    b. lose $30 and win $20?   **e.** find $5 and spend $5

    c. spend $28 and lose $15?

SECTION 2-6

1. All numbers to the left of zero on the number line are _____ numbers.

2. Name the directed number indicated by each letter on the number line shown below:

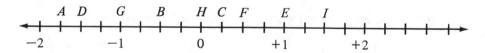

    a. *A*   **d.** *D*   **g.** *G*

    b. *B*   **e.** *E*   **h.** *H*

    c. *C*   **f.** *F*   **i.** *I*

3. Name the directed number(s) that is (are) described in each of the following:

    a. A number 15 units to the left of 0

    b. A number 3 units to the left of $+1$

    c. Two numbers, each 12 units from -4

    d. Two numbers, each 10 units from $+12$

4. A move from -9 to $+3$ is a move in the _____ direction.

5. Using a number line, express each change as a signed number.

a. From 0 to $+10$

d. From $+3$ to -1

b. From $+2$ to $+8$

e. From -4 to $+3$

c. From $+8$ to 0

f. From -2 to -7

6. Start at 0 and move 8 units in the positive direction; then move 5 units in the negative direction. Name the directed number at which you finish.

7. The temperature at 1 P.M. was $12°F$, but by 8 P.M. it had dropped 21 degrees. What was the temperature at 8 P.M.?

a. $33°F$ **b.** $-33°F$ **c.** $9°F$ **d.** $-9°F$

8. One day in February the temperature reading at 7:30 A.M. was $20°$ below zero. Later that day, the same thermometer had a reading of $5°$ above zero. What was the change in temperature between the two readings?

a. $25°$ **b.** $20°$ **c.** $15°$ **d.** $10°$

9. Name the opposite of:

a. 7 **b.** -3 **c.** 0 **d.** $\dfrac{1}{2}$ **e.** $-1\dfrac{1}{2}$

SECTION 2-7

1. TRUE or FALSE?

a. Each negative number is less than any positive number.

b. 0 is greater than every negative number.

c. $6 > 1$ **f.** $4 = -4$ **i.** $3 > -3$ **l.** $-5 < 3$

d. $0 > -3$ **g.** $-15 > -12$ **j.** $8 < 19$ **m.** $-3 > -5$

e. $4 < -1$ **h.** $-14 < -4$ **k.** $-3 > 5$ **n.** $-5 > -3$

2. Which number on the horizontal number line is associated with the point farther to the right?

a. $+3$ or -7 **b.** -4 or 0 **c.** -8 or -3 **d.** 0 or $+9$

3. Select the smaller of the two given numbers.

a. $+8$ or $+3$ **b.** -11 or -6 **c.** $+12$ or 0 **d.** 0 or -10 **e.** 6 or -8

4. Select the largest number in each of the following:

a. $8, 4, -10$ **c.** $0, 6, 1$ **e.** $2, -4, -7$ **g.** $-6, -8, -5$

b. $3, 7, 5$ **d.** $4, 0, -2$ **f.** $-14, 8, -3$ **h.** $-9, -11, -4$

5. Name the following groups of numbers in order of size with smallest first:

    a. $0, +17, -2, +8$ **b.** $+7, -3, -8, +5, -6$ **c.** $0, -16, 21, -21, 2$

6. For each of the following group of fractions:

    a. Find the LCD for each group.

    b. Rename each fraction in the group with the same denominator.

    c. Arrange the fractions in each group from smallest to largest.

    (1) $\dfrac{1}{2}, \dfrac{7}{8}$ (4) $\dfrac{3}{10}, \dfrac{1}{5}, \dfrac{9}{20}$

    (2) $\dfrac{-1}{3}, \dfrac{2}{9}$ (5) $\dfrac{3}{8}, \dfrac{1}{2}, \dfrac{-3}{16}$

    (3) $\dfrac{-3}{5}, \dfrac{-4}{15}$ (6) $\dfrac{3}{4}, \dfrac{-11}{8}, \dfrac{9}{16}, \dfrac{-33}{16}$

7. Insert the symbol "$<$" or "$>$" in each of the following so that each resulting statement is true.

    a. $-10 \underline{\quad} +10$ **b.** $-3 \underline{\quad} 0$ **c.** $3\dfrac{1}{3} \underline{\quad} 3\dfrac{2}{9}$ **d.** $-2\dfrac{5}{12} \underline{\quad} -2\dfrac{11}{24}$

SECTION 2-8

1. Find the least common multiple for each of the following pairs:

    a. 3 and 5   **d.** 5 and 8

    b. 10 and 12   **e.** 25 and 20

    c. 16 and 18   **f.** 24 and 16

2. Make each statement true by inserting the symbol "$=$," "$>$," or "$<$."

    a. $\dfrac{1}{2} \underline{\quad} \dfrac{2}{4}$ **g.** $\dfrac{1}{3} \underline{\quad} \dfrac{1}{4}$ **m.** $\dfrac{4}{7} \underline{\quad} \dfrac{4}{9}$ **s.** $\dfrac{1}{8} \underline{\quad} \dfrac{1}{12}$

    b. $\dfrac{5}{4} \underline{\quad} \dfrac{4}{3}$ **h.** $\dfrac{1}{3} \underline{\quad} \dfrac{-3}{7}$ **n.** $\dfrac{-5}{6} \underline{\quad} \dfrac{-2}{5}$ **t.** $\dfrac{31}{3} \underline{\quad} \dfrac{51}{5}$

    c. $\dfrac{5}{9} \underline{\quad} \dfrac{5}{10}$ **i.** $\dfrac{2}{7} \underline{\quad} \dfrac{-1}{4}$ **o.** $\dfrac{6}{9} \underline{\quad} \dfrac{10}{15}$ **u.** $2\dfrac{1}{3} \underline{\quad} 2\dfrac{1}{5}$

    d. $\dfrac{0}{7} \underline{\quad} \dfrac{0}{6}$ **j.** $\dfrac{-3}{4} \underline{\quad} \dfrac{-1}{10}$ **p.** $\dfrac{5}{7} \underline{\quad} \dfrac{3}{4}$ **v.** $-4\dfrac{1}{3} \underline{\quad} -4\dfrac{1}{4}$

    e. $\dfrac{199}{200} \underline{\quad} \dfrac{8}{8}$ **k.** $\dfrac{-5}{8} \underline{\quad} \dfrac{-5}{7}$ **q.** $\dfrac{-5}{7} \underline{\quad} \dfrac{-3}{4}$ **w.** $-7\dfrac{3}{4} \underline{\quad} -7\dfrac{1}{2}$

    f. $\dfrac{3}{3} \underline{\quad} \dfrac{8}{8}$ **l.** $\dfrac{3}{3} \underline{\quad} \dfrac{3}{5}$ **r.** $\dfrac{3}{4} \underline{\quad} \dfrac{2}{3}$ **x.** $\dfrac{-5}{4} \underline{\quad} \dfrac{-6}{5}$

SECTION 2-9

1. Find the least common multiple for each of the following groups of numbers:

    a. 2, 5, and 8   **c.** 12, 18, and 30   **e.** 8, 10, and 12

    b. 4, 8, and 16   **d.** 14, 21, and 35   **f.** 15, 40 and 75

2. Arrange in order from least to greatest:

    a. $\dfrac{3}{5}, \dfrac{2}{3}, \dfrac{7}{10}$ **e.** $\dfrac{-3}{10}, \dfrac{-9}{50}, \dfrac{-7}{25}$ **i.** $\dfrac{1}{20}, \dfrac{1}{30}, \dfrac{1}{60}$

    b. $\dfrac{3}{4}, \dfrac{11}{8}, \dfrac{9}{16}, \dfrac{33}{16}$ **f.** $\dfrac{-1}{7}, \dfrac{-3}{14}, \dfrac{-5}{6}$ **j.** $4\dfrac{1}{4}, 4\dfrac{1}{3}, 4\dfrac{5}{12}$

    c. $\dfrac{4}{10}, \dfrac{-2}{3}, \dfrac{-3}{5}$ **g.** $\dfrac{3}{7}, \dfrac{3}{3}, \dfrac{3}{9}$ **k.** $-3\dfrac{1}{8}, +3\dfrac{5}{12}, +3\dfrac{5}{6}$

    d. $\dfrac{7}{18}, \dfrac{30}{36}, \dfrac{-3}{4}$ **h.** $\dfrac{2}{3}, \dfrac{2}{12}, \dfrac{2}{11}$ **l.** $-1\dfrac{2}{3}, -1\dfrac{3}{5}, -1\dfrac{1}{2}$

3. Which fraction is largest in value?

    a. $\dfrac{2}{3}$ **b.** $\dfrac{5}{8}$ **c.** $\dfrac{3}{5}$ **d.** $\dfrac{1}{2}$

4. Which fraction is smallest in value?

    a. $\dfrac{1}{9}$ **b.** $\dfrac{1}{12}$ **c.** $\dfrac{1}{19}$ **d.** $\dfrac{1}{23}$

SECTION 2-10

1. Complete so that the resulting statement is true.

    a. The sum of two positive integers is a _____ integer.

    b. The sum of two negative integers is a _____ integer.

2. Which number sentence is shown by the arrows along the number line below?

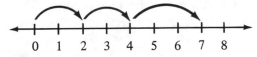

    0   1   2   3   4   5   6   7   8

3. Use the number line to find the sums.

    a. $(+5) + (+13)$ **g.** $(-6) + (+4)$ **m.** $+5 + (-7)$

    b. $(-11) + (-8)$ **h.** $8 + (-2)$ **n.** $6 + (-3)$

    c. $(-6) + (-21)$ **i.** $-3 + 14$ **o.** $-6 + 3$

    d. $(-12) + (-3)$ **j.** $6 + (-8)$ **p.** $-7\dfrac{1}{2} + 9$

    e. $(0) + (+4)$ **k.** $-9 + 17$ **q.** $11 + \left(-16\dfrac{1}{4}\right)$

    f. $(0) + (-3)$ **l.** $-5 + 2$ **r.** $-12 + 6\dfrac{3}{4}$

SECTION 2-11

1. Find the absolute value of:

   a. -4 **b.** $+5$ **c.** $\dfrac{-1}{3}$ **d.** $+2\dfrac{1}{2}$ **e.** $-3\dfrac{1}{5}$

2. Which is greater, the absolute value of $+7$ or the absolute value of -8?

3. Which has the greater absolute value, -9 or $+7$?

4. Find the additive inverse of:

   a. $+5$ **b.** -2 **c.** $\dfrac{-1}{4}$ **d.** $+1\dfrac{1}{8}$

5. Which has the greater opposite number, -5 or $+10$?

6. Write an integer for each sum.

a. $(-13) + (-9)$	**g.** $-6 + (-2)$	**m.** $6 + (-2)$
b. $(-15) + (-16)$	**h.** $(-4) + 2$	**n.** $5 + (-12)$
c. $(-48) + (-57)$	**i.** $-1 + 6$	**o.** $11 + (-11)$
d. $(+39) + (+78)$	**j.** $-7 + (-3)$	**p.** $-23 + 18$
e. $(+72) + (+49)$	**k.** $-9 + (+9)$	**q.** $-60 + 40$
f. $(+26) + (37)$	**l.** $-6 + (+11)$	**r.** $(+6) + (-2) + (+7)$

   s. $(-8) + (-6) + (+15) + (-2)$ **v.** $(-14) + 6 + 10 + (-12)$

   t. $(-12) + (-17) + (-12) + (-8)$ **w.** $(-30) + 12 + (-26)$

   u. $(-12) + 5 + 15 + (-18)$ **x.** $(-9) + 7 + (-5) + 8$

7. A man makes a deposit of $1200 in his checking account. Then he withdraws $950. What is the net change in his checking account balance?

8. A football team gained 11 yards, lost 13 yards, and lost 7 yards. What was the net result?

9. Start at 20° below zero, up 30°, down 20°, up 40°, up 2°, down 15°, up 30°, down 52°. What is the final temperature?

10. Tom had $25 on Friday and owed $12 on Monday. How much did he spend over the weekend?

11. Evaluate each of the following if $x = -2$, $y = -1$, and $z = +3$:

    a. $x + y$ **b.** $y + z$ **c.** $x + z$ **d.** $x + y + z$

SECTION 2-12

1. Find the sum for each of the following:

a. $\left(\dfrac{-1}{2}\right) + \left(\dfrac{-1}{2}\right)$ e. $\left(\dfrac{-3}{16}\right) + \left(\dfrac{-9}{16}\right)$ i. $\left(+10\dfrac{1}{2}\right) + \left(7\dfrac{1}{2}\right)$

b. $\left(\dfrac{-3}{5}\right) + \left(\dfrac{-1}{5}\right)$ f. $\left(-1\dfrac{1}{3}\right) + \left(\dfrac{-2}{3}\right)$ j. $\left(+69\dfrac{1}{3}\right) + \left(+17\dfrac{1}{3}\right)$

c. $\left(\dfrac{+20}{10}\right) + \left(\dfrac{+32}{10}\right)$ g. $\left(\dfrac{+4}{7}\right) + \left(9\dfrac{5}{7}\right)$ k. $\left(-2\dfrac{3}{4}\right) + \left(-4\dfrac{1}{4}\right)$

d. $\left(\dfrac{-28}{4}\right) + \left(\dfrac{-3}{4}\right)$ h. $\left(+5\dfrac{1}{8}\right) + \left(+4\dfrac{3}{8}\right)$ l. $\left(-2\dfrac{1}{2}\right) + \left(-1\dfrac{1}{2}\right)$

m. $\left(-3\dfrac{3}{4}\right) + \left(-2\dfrac{1}{4}\right) + \left(-1\dfrac{3}{4}\right) + \dfrac{-1}{4}$

n. $\left(-5\dfrac{1}{4}\right) + \left(-2\dfrac{3}{4}\right) + \left(-7\dfrac{3}{4}\right) + \left(-4\dfrac{1}{4}\right)$

o. $\left(\dfrac{+3}{5}\right) + \left(\dfrac{+7}{10}\right)$ t. $\left(\dfrac{-2}{3}\right) + \left(\dfrac{-3}{4}\right)$ y. $\dfrac{+1}{5} + \dfrac{+2}{3} + \dfrac{+4}{7}$

p. $\left(\dfrac{-1}{8}\right) + \left(\dfrac{-3}{4}\right)$ u. $\left(\dfrac{-3}{4}\right) + \left(\dfrac{-7}{8}\right)$ z. $\left(-5\dfrac{5}{6}\right) + \left(-2\dfrac{2}{3}\right)$

q. $\left(\dfrac{+5}{6}\right) + \left(\dfrac{+3}{4}\right)$ v. $\left(\dfrac{-7}{12}\right) + \left(\dfrac{-5}{6}\right)$ aa. $\left(-1\dfrac{7}{8}\right) + \left(-2\dfrac{2}{3}\right)$

r. $\left(\dfrac{-3}{4}\right) + \left(\dfrac{-1}{2}\right)$ w. $\left(\dfrac{-1}{5}\right) + \left(\dfrac{-1}{3}\right)$ bb. $\left(+3\dfrac{9}{10}\right) + \left(10\dfrac{5}{8}\right)$

s. $\left(\dfrac{+2}{3}\right) + \left(\dfrac{+3}{4}\right)$ x. $\dfrac{1}{6} + \dfrac{2}{3} + \dfrac{1}{8}$

2. Find the total number of yards in three rolls of cloth if their measurements in yards are $24\dfrac{3}{8}$, $24\dfrac{7}{16}$, and $24\dfrac{5}{8}$.

3. Ron, Clyde, and Jack went fishing. Ron's catch weighed $3\dfrac{1}{4}$ pounds. Clyde's catch weighed $5\dfrac{3}{7}$ pounds. Jack's catch weighed $4\dfrac{1}{6}$ pounds. What was the total weight?

SECTION 2-13

1. Find the sum for each of the following:

a. $\dfrac{1}{2} + \dfrac{-1}{4}$ d. $\dfrac{-7}{8} + \dfrac{1}{4}$ g. $\dfrac{11}{18} + \dfrac{-11}{18}$ j. $-2 + 3\dfrac{1}{2}$

b. $\dfrac{3}{4} + \dfrac{-5}{8}$ e. $\dfrac{7}{10} + \dfrac{-9}{10}$ h. $\dfrac{-7}{3} + \dfrac{4}{3}$ k. $-3\dfrac{1}{8} + \dfrac{+24}{8}$

c. $\dfrac{7}{8} - \dfrac{1}{3}$ f. $\dfrac{-2}{5} + \dfrac{3}{5}$ i. $\dfrac{2}{5} + \dfrac{-3}{10}$ l. $+4\dfrac{3}{8} + -5\dfrac{3}{8}$

m. $-2\dfrac{2}{3} + +3\dfrac{1}{3}$

n. $\dfrac{-1}{2} + \dfrac{-4}{5} + \dfrac{+3}{2} + \dfrac{+9}{5}$

o. $\dfrac{+2}{5} + \dfrac{-5}{6} + \dfrac{+1}{6} + \dfrac{-7}{15}$

p. $+6\dfrac{1}{4} + -3\dfrac{1}{2}$

q. $\left(3\dfrac{1}{4}\right) + \left(+8\dfrac{1}{2}\right) + \left(-40\dfrac{3}{4}\right)$

r. $\left(+8\dfrac{3}{5}\right) + \left(-2\dfrac{7}{10}\right)$

s. $\left(+2\dfrac{1}{2}\right) + \left(-1\dfrac{1}{4}\right) + \left(-5\dfrac{3}{4}\right) + (+7)$

2. Perform the subtraction indicated on each of the following arithmetic numbers:

a. $\dfrac{8}{5} - \dfrac{3}{5}$

b. $\dfrac{5}{16} - \dfrac{2}{16}$

c. $\dfrac{3}{4} - \dfrac{1}{6}$

d. $\dfrac{3}{14} - \dfrac{4}{21}$

e. $7\dfrac{3}{4} - 2\dfrac{1}{4}$

f. $8\dfrac{1}{3} - 5\dfrac{2}{3}$

g. $6\dfrac{1}{7} - 4\dfrac{3}{14}$

h. $8\dfrac{1}{3} - 6\dfrac{1}{2}$

i. $9 - \dfrac{5}{11}$

j. $26 - 3\dfrac{4}{9}$

Measuring Your Vocabulary

Column II contains the meanings or descriptions of the terms or symbols in Column I, which are used in this chapter.

For each number from Column I, write the letter from Column II that corresponds to the best meaning or description of the term or symbol.

Column I	**Column II**
1. Mixed number	a. Set of numbers including 0, 1, 2, ...
2. Fraction	b. A number that shows both quantity and direction.
3. Integers	
4. Improper fraction	c. A fraction in which the numerator is equal to or greater than the denominator.
5. Whole number	
6. Line segment	d. A mathematical sentence that uses the symbol "<" or ">."
7. Signed number	
8. Line	e. The smallest number that is a multiple of 2 or more numbers.
9. >	
10. Equivalent fractions	f. The quotient of two quantities, denominator not zero.
11. Equation	
12. Number line	g. Set of numbers including 0, positive whole numbers, and negative whole numbers.
13. Proper fraction	
14. Negative number	h. The symbol that means "is less than."
15. Directed numbers	i. A mathematical sentence that uses the symbol "=."
16. Inequality	
17. Like fractions	j. A part of a straight line.
18. Absolute value	k. A straight line on which points are associated with numbers.
19. <	
20. Numbers that are opposites	l. Numbers that are the same distance away from 0 on number line, but are on opposite sides from 0.
21. Least common multiple	
22. Positive numbers	m. A whole number plus a fraction.
23. LCD	n. Two fractions that name the same number.
	o. A geometric figure that contains a set of points but no endpoints.
	p. The symbol that means "is greater than."
	q. The symbol that means "is equal to."
	r. A fraction whose numerator is less than its denominator.
	s. Fractions that have equal denominators.
	t. The smallest number that each of the given denominators will divide without a remainder.
	u. Fractions with unequal denominators.
	v. Numbers to the left of 0 on the number line.
	w. The number of units that a number is represented from 0 on the number line without regard to direction.
	x. Numbers to the right of 0 on the number line.
	y. The positive and negative numbers.
	z. A part of a straight line that has a single endpoint.

Chapter 3
Metric Measurement

3-1 *Units of Linear Measurement*

STUDY GUIDE Scientists throughout the world measure lengths, distances, weights, and other values by a *standard* system called the *metric system*. The basic unit of length in the metric system is the *meter*, which is a little longer than a yard.

A group of French scientists developed the metric system, and it became the legal system of weights and measures in France almost 200 years ago. Most countries throughout the world now use the metric system in daily life for measuring lengths, weighing objects, and making other measurements. The United States is moving toward adoption of the metric system for all everyday measurements.

In the English system, the basic unit of length is the *yard*. The yard is divided into smaller units (feet and inches) and is multiplied to build larger units (miles). The yard is exactly 36 inches long or exactly 3 feet long. Since 12 inches = 1 foot, 36 inches = 3 feet = 1 yard.

In the metric system, the basic unit of length is the *meter*. The meter is also divided into smaller units (decimeters, centimeters, and millimeters) and is multiplied to build larger units (dekameters, hectometers, and kilometers) for different measurements of lengths and distances. The meter is exactly 10 decimeters long or exactly 100 centimeters long or exactly 1000 millimeters long. In comparison with units of *linear measure* in the English system, the meter is *approximately* 39.37 inches long. Compare:

YARD STICK—36 inches (exactly)	English system

YARD STICK—3 feet (exactly)	English system

METER STICK—100 centimeters (exactly)	Metric system

METER STICK—1000 millimeters (exactly)	Metric system

METER STICK—39.37 inches (approx.)	Comparing metric with English

The following table shows some of the metric units of linear measure:

Unit	Abbreviation	Length	Unit	Abbreviation	Length
kilometer	km	1000 meters	decimeter	dm	$\frac{1}{10}$ meter
hectometer	hm	100 meters			
dekameter	dkm	10 meters	centimeter	cm	$\frac{1}{100}$ meter
meter	m	1 meter			
			millimeter	mm	$\frac{1}{1000}$ meter

Notice that all units are obtained from the meter by multiplying or dividing by 10 or 100 or 1000. These numbers are "powers of 10." Study the following illustrations of metric units of length:

1 mm   1 cm

1 decimeter (1 dm)

Some examples of the use of metric units are as follows: a film is 35 millimeters wide; a pencil is 20 centimeters long; a basketball player is 2 meters tall; California is 4800 kilometers from New York.

The most common units of length are the *kilometer, meter, centimeter,* and *millimeter*. The kilometer is a large unit for measuring long distances, the meter is a midsize unit, and the centimeter and millimeter are small units. The abbreviations for these units are not followed by periods.

FACT FINDING

1. The metric system was developed in _____ about
_____ years ago.

2. It is used by _____ everywhere in the world.

3. The basic unit of linear measure in the metric system is the
_____ .

4. The meter is _____ (<, =, or >) the yard.

5. The meter is divided into _____ units and multiplied to make
_____ units.

6. All units are obtained from the meter by multiplying or dividing by
powers of _____ .

7. The most common units of length are the _____ ,
_____ , _____ , and _____ .

8. The large unit of linear measure is the _____ .

9. The small units of linear measure are the _____ and the
_____ .

10. Abbreviations for metric units of linear measure are not followed by
_____ .

MODEL PROBLEMS

1. A baseball player hit a home run over the fence, 120 meters from home plate. He then ran around the bases for a total distance of 120 yards. Which is the greater distance?

   Solution: A meter is longer than a yard, so the ball flew through the air a greater distance than the player ran around the bases.

2. Match each item with the appropriate unit for measuring its length:

   (1)   Distance from New York to Florida   a.   mm

   (2)   Length of a piece of chalk   b.   cm

   (3)   Width of movie film   c.   m

   (4)   Height of a room   d.   km

   Solution: (1) d   (2) b   (3) a   (4) c

EXERCISES

1. The following words or phrases were in *italics* in this section. Look up their definitions in a dictionary and write their meanings.

a. Standard	**e.** Approximately
b. Metric system	**f.** Kilometer
c. Meter	**g.** Centimeter
d. Linear measure	**h.** Millimeter

2. Match the number with the letter.

   a.   Centimeter   (1)   1000 meters

   b.   Millimeter   (2) $\frac{1}{100}$ meter

   c.   Kilometer   (3) $\frac{1}{1000}$ meter

3. Place the symbol "<" or ">" between each pair of units to make the statement true.

   a.   Millimeter __ kilometer   **d.**   Meter __ kilometer

   b.   Centimeter __ meter   **e.**   Centimeter __ millimeter

   c.   Meter __ millimeter   **f.**   Kilometer __ centimeter

4. A football field is 100 yards long from goal post to goal post. A race in the Olympic Games is 100 meters long. Which is the greater distance?

5. In the words millimeter, centimeter, and kilometer, the underlined parts are called "prefixes." Look up the word "prefix" in a dictionary and write its definition.

6. Arrange these units of length in order from smallest to largest:

   a.   Meter   **b.**   Centimeter   **c.**   Kilometer   **d.**   Millimeter

7. For each of the following, write the abbreviation for the unit that would probably be used to measure the length, width, or distance. Select units from the following: mm, cm, m, km.

a. Width of a notebook

b. Distance of a bus or train ride

c. Length of a giraffe's neck

d. Length of a pencil point

e. Height of the Empire State Building

f. Height of a pupil

g. Height of a flagpole

h. Distance from Boston to Washington

i. Distance of a marathon race

j. Width of the pupil in your eye

k. Thickness of a dime

l. Length of a new pencil

m. Length of your thumb

n. Width of a fingernail

o. Distance a plane can fly in one hour

p. Distance across the Atlantic Ocean

q. Width of your desk

r. Length of an eyelash

s. Height of an elephant

t. Height of Niagara Falls

**EXTRA
FOR
EXPERTS**

If 1000 meters is added to 100 dekameters and 10 hectometers, what is the total length in kilometers?

3-2 *Large Numbers and Decimals*

STUDY GUIDE Metric units of linear measure, such as the millimeter, centimeter, meter, and kilometer, can be converted from one to another by multiplying or dividing by 10 or 100 or 1000, and so on. These numbers are *powers of 10.* Our number system is called a *decimal system* because it is based on powers of 10. When we write *numerals,* each *digit* in any numeral has a value depending on its place in the numeral. This is called *place value.* Study the following diagram:

100,000	10,000	1000	100	10	1	$\frac{1}{10}$ or 0.1	$\frac{1}{100}$ or 0.01	$\frac{1}{1000}$ or 0.001
1 followed by 5 zeros	1 followed by 4 zeros	1 followed by 3 zeros	1 followed by 2 zeros	1 followed by 1 zero	1 followed by 0 zero	1 in the last of 1 decimal place	1 in the last of 2 decimal places	1 in the last of 3 decimal places
$\longleftarrow$ 10^5	10^4	10^3	10^2	10^1	10^0	10^{-1}	10^{-2}	10^{-3} $\longrightarrow$
hundred thousand	ten thousand	thousand	hundred	ten	unit	tenth	hundredth	thousandth

The diagram may be extended to the left for larger numbers (for example, $1,000,000 = 10^6$ is the million place) and to the right for smaller numbers (for example, $\frac{1}{10,000} = 0.0001 = 10^{-4}$ is the ten-thousandth place). It may be extended in either or both directions as far as we wish.

A decimal point is placed between the digit written in the unit place and the digit written in the tenth place to separate the whole-number part of the numeral from the decimal part. For example, 3,507,284.15 is read as "three million, five hundred seven thousand, two hundred eighty-four, and fifteen hundredths." Note that the digit 0 in the ten-thousand place means that there are no ten thousands in this number. The decimal point is always read as the word "and."

Note that $1000 = 10^3$, $1 = 10^0$, and $0.01 = 10^{-2}$. The numerals 3, 0, and -2 are called *exponents*. Exponents may be positive, zero, or negative numerals, and they may be used to express numerals as powers of 10. Positive and zero powers of 10 represent the whole-number part of numerals, while negative powers of 10 represent the decimal part of numerals.

FACT FINDING

1. Our number system is based on powers of _____. For this reason it is called a _____ system.

2. Each digit in a numeral has a _____ value.

3. Exponents may be positive numerals or _____ or _____ numerals.

4. A _____ point separates the whole-number part of a numeral from the _____ part.

5. Positive and zero powers of 10 represent the _____ part of a numeral, while negative powers of 10 represent the _____ part of a numeral.

6. Metric units of linear measure can be converted from one to another by _____ or _____ by powers of 10.

MODEL PROBLEMS

1. Write as a numeral: seventy-three thousand, four hundred sixty-nine.

Solution: 73,469.

2. Write in words: 389.07.

Solution: Three hundred eighty-nine and seven hundredths. Note that there are no tenths in this number.

3. What is the place value of each of the following digits in the numeral 17,564.389?

a. 1   b. 5   c. 9

Solution: a. Ten thousand   b. Hundred   c. Thousandth

1	7	5	6	4	.3	8	9
ten thousand	thousand	hundred	ten	unit	tenth	hundredth	thousandth

This numeral may be read as "seventeen thousand, five hundred sixty-four, and three hundred eighty-nine thousandths."

4. Express as a decimal: $\frac{37}{1000}$.

Solution: This fraction is read as "thirty-seven thousandths." Thousandths requires three places to the right of the decimal point. Therefore this is written in decimal form as 0.037. Zero is used as a place holder to fill the tenth place, since 37 has only two digits and thousandths requires three places to the right of the decimal point.

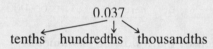

0.037
tenths   hundredths   thousandths

5. Express as a common fraction: 0.09.

Solution: Since there are two decimal places in this numeral, it is read as "nine hundredths." This may be written as the common fraction $\frac{9}{100}$.

EXERCISES

1. The following words or phrases were in *italics* in this section. Write the definition of each:

a. Powers of 10   d. Digit

b. Decimal system   e. Place value

c. Numerals   f. Exponents

2. Fill in the correct exponent in each of the following:

a. $1,000,000 = 10^?$ c. $100 = 10^?$ e. $0.1 = 10^?$

b. $10,000 = 10^?$ d. $1 = 10^?$ f. $0.001 = 10^?$

3. Write the numeral for each of the following:

a. $10^5 = $ __   c. $10^2 = $ __   e. $10^{-2} = $ __

b. $10^3 = $ __   d. $10^0 = $ __   f. $10^{-4} = $ __

4. Write each of the following numerals in symbols and in words:

   a. 5 in the thousand place, 3 in the hundred place, 9 in the ten place, 0 in the unit place, 6 in the tenth place.

   b. 7 in the million place, 4 in the thousand place, 8 in the ten place, and 0 in all other places.

   c. 3 in the tenth place, 8 in the hundredth place, 5 in the thousandth place.

   d. 3 in the ten place, 0 in the unit place, 0 in the tenth place, 1 in the hundredth place.

   e. 7 in the unit place, 0 in the tenth place, 2 in the hundredth place, 1 in the thousandth place, 1 in the ten-thousandth place.

5. What is the value of the 4 in each of the following numerals?

   a. 40,000   b. 2431   c. 348   d. 64   e. 5.4   f. 0.004

6. Write each of the following in words:

   a. 32.8   c. 506.72   e. 5650   g. 0.509   i. 0.2
   b. 105   d. 7.13   f. 65,000   h. 100,000 j. 3,000,000

7. Write each of the following in numerals:

   a. Thirteen and five tenths   f. Five hundred three

   b. Four hundred thousand   g. Sixty-seven and four hundredths

   c. Eighteen hundredths   h. Ten thousand

   d. Three thousandths   i. Two hundredths

   e. One million, two hundred thousand   j. One and three tenths

8. Express each of the following as a decimal:

   a. $\dfrac{7}{10}$ c. $\dfrac{19}{100}$ e. $\dfrac{13}{1000}$ g. $\dfrac{65}{10,000}$

   b. $\dfrac{3}{100}$ d. $\dfrac{9}{1000}$ f. $\dfrac{603}{1000}$

9. Express each of the following as a common fraction:

   a. 0.8   c. 0.23   e. 0.053   g. 0.1057
   b. 0.03   d. 0.007 f. 0.923

10. A radio announcement indicated that the barometric pressure was 30.17. Write this numeral in words.

11. A student's height was measured as one and sixty-seven hundredths meters. Write this in numerals.

12. The height of a flagpole was 15.3 meters. Write this numeral in words.

13. A car traveled twenty-eight and three tenths kilometers. Write this in numerals.

14. Write 3.14 cm in words.

15. Write two and seven tenths millimeters in numerals.

**EXTRA
FOR
EXPERTS**

Arrange the following in order from smallest to largest:

$$\frac{2}{10} \qquad 0.043 \qquad \frac{5}{100} \qquad 0.194 \qquad \frac{6}{1000} \qquad 0.8 \qquad \frac{17}{100}$$

3-3 *The Metric Ruler: Meters and Centimeters*

STUDY GUIDE A meter stick is 1 meter long. Just as a yardstick is divided into feet and inches, a meter stick is divided into 100 smaller units, called *centimeters*, which can be used to measure the lengths of smaller objects. To measure the lengths of *very* small objects, centimeters can be divided into 10 equal parts. Each part is then $\frac{1}{10}$ or 0.1 centimeter.

This small length is called a *millimeter*. Millimeter lengths will be studied in the next section.

Since 1 meter = 100 centimeters, it is possible to change lengths expressed in meters to centimeters by *multiplying by 100*. One hundred is a power of 10: $100 = 10^2$.

Multiplication of a number by a power of 10 can be done quickly by moving the decimal point in the given number to the right the same number of places as the exponent.

Examples: $6.431 \quad \times 10^2 = 6.43.1$

$.8975 \quad \times 10^3 = .897.5$

$78.16342 \times 10^3 = 78.163.42$

Multiplication of any number by 100 can be done quickly by moving the decimal point in the number *two places to the right*. For whole numbers, this requires placing two zeros at the right-hand end of the number. Study the table on page 122.

$$0.01 \text{ m} = 1 \text{ cm}$$
$$0.02 \text{ m} = 2 \text{ cm}$$
$$0.03 \text{ m} = 3 \text{ cm}$$
$$\text{etc.}$$
$$0.09 \text{ m} = 9 \text{ cm}$$
$$0.10 \text{ m} = 10 \text{ cm} \qquad \text{or} \qquad 0.1 \text{ m} = 10 \text{ cm}$$
$$0.2 \text{ m} = 20 \text{ cm}$$
$$0.3 \text{ m} = 30 \text{ cm}$$
$$\text{etc.}$$
$$0.9 \text{ m} = 90 \text{ cm}$$
$$1.0 \text{ m} = 100 \text{ cm}$$

Note that $0.10 = 0.1$ since $\dfrac{10}{100} = \dfrac{1}{10}$.

FACT FINDING

1. A meter stick can be divided into 100 smaller units called _____ .

2. One hundredth of a meter is called a _____ .

3. One meter = _____ centimeters.

4. One tenth of a centimeter is called a _____ .

5. Lengths expressed in meters can be changed to centimeters by

    _____ by _____ .

MODEL PROBLEM

One pupil in a class is 1.5 meters tall, and another pupil is 155 centimeters tall. Which pupil is taller?

Solution: Meters can be changed to centimeters by multiplying by 100. Therefore 1.5 meters = 1.5 × 100 cm = 150 centimeters tall. The pupil who is 155 centimeters tall is the taller pupil. Note that, since $100 = 10^2$, the decimal point in 1.5 was moved two places to the right. In order to do this, it was necessary to place a zero at the right end of the numeral: 1.5 × 100 = 1.50. = 150.

EXERCISES

1. Change each measurement from meters to centimeters:

    a. 3 m   **c.** 138 m   **e.** 16.1 m   **g.** 67.29 m   **i.** 0.08 m

    b. 27 m   **d.** 8.3 m   **f.** 5.83 m   **h.** 0.4 m   **j.** 0.005 m

2. Study the part of the metric ruler shown:

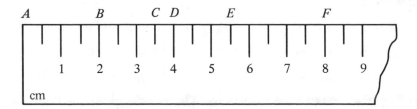

Find the distance between each of the following pairs of points:

a. *A* and *B*   **c.** *A* and *D*   **e.** *B* and *D*   **g.** *E* and *F*

b. *A* and *C*   **d.** *B* and *C*   **f.** *C* and *D*

3. Use a metric ruler to find the length of each of the following line segments to the nearest tenth of a centimeter:

a. _____

b. _____

c. _____

d. _____

e. _____

4. The length of a yardstick is approximately 0.9 meter. Express this length in centimeters.

5. Measure the length of each of the following with a metric ruler or meter stick, and record your answers in the units shown:

a. Your height in cm

b. The length of the classroom in m

c. The length of your thumb in cm

d. The thickness of your notebook in cm or to the nearest tenth of a cm

e. The width of your desk in cm

f. The length of one of the hairs from your head in cm

g. The width of a dime to the nearest tenth of a cm

h. The thickness of a quarter to the nearest tenth of a cm

i. The length of the teacher's desk in cm or to the nearest tenth of a m

j. Your waist measurement in cm (use a string and place it alongside a metric ruler if you cannot get a metric tape measure)

6. Place the correct symbol (<, =, or >) between each pair of measurements.

    a. 7 m ___ 750 cm   **f.** 16 m ___ 1600 cm

    b. 5.3 m ___ 500 cm   **g.** 1.08 m ___ 108 cm

    c. 3.9 m ___ 309 cm   **h.** 0.3 m ___ 30 cm

    d. 15.7 m ___ 157 cm   **i.** 0.05 m ___ 50 cm

    e. 32 m ___ 320 cm   **j.** 0.007 m ___ 7 cm

7. How many centimeters is $\frac{1}{2}$ meter? $\frac{1}{4}$ meter? $\frac{3}{4}$ meter?

 **EXTRA FOR EXPERTS**

A person is 1.95 m tall. He buys a pair of cross-country skis that are 210 cm long. He also buys ski poles (for pushing himself along on level snow) that are $\frac{2}{3}$ the length of his skis.

    a. How much longer are the skis than the height of the skier?

    b. How many centimeters long are the ski poles?

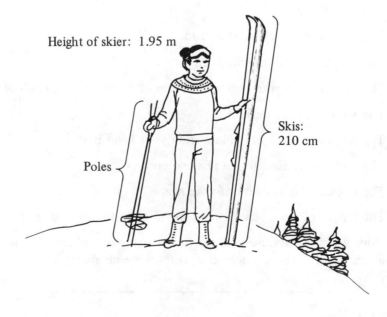

Height of skier: 1.95 m

Skis: 210 cm

Poles

3-4 *Comparing Centimeters and Millimeters; Comparing Meters and Millimeters*

STUDY GUIDE One tenth of a centimeter is called a *millimeter*. A millimeter is a small unit for measuring length. The prefix "milli-" means one thousandth, which, written in numerals, is 0.001 or $\frac{1}{1000}$.

Therefore 1 millimeter = $\frac{1}{1000}$ meter or 0.001 meter. Study the following table:

```
0.1 cm =  1 mm
0.2 cm =  2 mm
0.3 cm =  3 mm
        etc.
0.9 cm =  9 mm
1.0 cm = 10 mm    or    1 cm =   10 mm
                        2 cm =   20 mm
                        3 cm =   30 mm
                              etc.
                        9 cm =   90 mm
                       10 cm =  100 mm    so    10 cm =   100 mm
                                                20 cm =   200 mm
                                                30 cm =   300 mm
                                                      etc.
                                                90 cm =   900 mm
                                               100 cm = 1000 mm = 1 m
```

Note that 1 cm = 10 mm and 1 m = 100 cm = 1000 mm.

Centimeters can be changed to millimeters by *multiplying by 10*.
Meters can be changed to centimeters by *multiplying by 100*.
Meters can be changed to millimeters by *multiplying by 1000*.

Note that it is always possible to change from larger units to smaller units of measure by *multiplying by an appropriate power of 10*. Recall that $10 = 10^1$, $100 = 10^2$, and $1000 = 10^3$. These are powers of 10. Multiplication by powers of 10 can be done quickly by moving the decimal point to the right as many places as are indicated by the exponent.

FACT FINDING

1. A millimeter is _____ _____ of a centimeter.

2. The prefix "milli-" means _____.

3. One centimeter = _____ millimeters.

4. One meter = _____ centimeters = _____ millimeters.

5. Centimeters can be changed to millimeters by _____ by 10.

6. Meters can be changed to millimeters by _____ _____ _____ .

7. Larger units of measurement can be changed to smaller units by multiplying by _____ of 10.

MODEL PROBLEMS

1. The length of a ballpoint pen is 15 cm. Express this length in mm.

   Solution: Since cm can be changed to mm by multiplying by 10,
   $15 \times 10 = 150$ mm.

2. Change 3.25 m to mm.

   Solution: Since m can be changed to mm by multiplying by 1000,
   $3.25 \times 1000 = 3250$ mm.
   Note that, since $1000 = 10^3$, the decimal point in 3.25 was moved three places to the right. In order to do this, it was necessary to place a zero at the right end of the numeral: $3.25 \times 1000 = 3.250 = 3250$.

EXERCISES

1. Convert each of the following measurements in centimeters to millimeters:

a. 7 cm	**c.** 137 cm	**e.** 0.48 cm	**g.** 18.06 cm	**i.** 72.4 cm
b. 50 cm	**d.** 0.7 cm	**f.** 3.9 cm	**h.** 9.01 cm	**j.** 3.14 cm

2. Convert each of the following measurements in meters to millimeters:

a. 5 m	**c.** 250 m	**e.** 0.8 m	**g.** 7.05 m	**i.** 14.7 m
b. 63 m	**d.** 0.38 m	**f.** 45.6 m	**h.** 6.1 m	**j.** 63.25 m

3. Convert each of the following measurements to the unit indicated:

a. 13 m = __ cm	**d.** 3.8 m = __ cm	**g.** 0.67 m = __ cm
b. 4.5 m = __ mm	**e.** 27 m = __ mm	**h.** 0.5 m = __ mm
c. 0.52 cm = __ mm	**f.** 0.4 cm = __ mm	**i.** 9.2 cm = __ mm

4. A penny has a thickness of approximately 0.1 cm. Express this measurement in millimeters.

5. A roll of nickels contains 40 nickels. If the thickness of one nickel is approximately 0.2 cm, calculate the height of a roll of nickels in centimeters. Convert the result to millimeters.

6. Use a metric ruler to draw a line segment having a length of:

a. 8 cm	**c.** 5.3 cm	**e.** 12.5 cm	**g.** 1 cm	**i.** 2.5 cm
b. 47 mm	**d.** 85 mm	**f.** 7 mm	**h.** 100 mm	**j.** 35 mm

7. The length of a yardstick is approximately 0.9 meter. Express this length in millimeters.

8. Measure the length of each of the following with a metric ruler and record the answers in millimeters:

    a. The distance between the lines on a page in your notebook

    b. The thickness of a dime

    c. The width of a quarter

    d. The length of your thumbnail

    e. The length of a straight pin

    f. The length of a piece of chalk

    g. The thickness of this book

    h. The width of a board eraser

    i. The length of the eraser on a pencil

    j. The length of a pencil point

9. Place the correct symbol ($<$, $=$, or $>$) between each pair of measurements.

a. 37 cm __ 370 mm	**f.** 14.3 cm __ 1430 mm
b. 5.8 m __ 580 mm	**g.** 27 m __ 270 mm
c. 6.3 cm __ 603 mm	**h.** 63 cm __ 6300 mm
d. 15 m __ 1500 cm	**i.** 1 m __ 1000 mm
e. 5 m __ 5000 mm	**j.** 100 cm __ 1000 mm

10. How many millimeters is $\frac{1}{2}$ centimeter? $\frac{1}{2}$ meter?

A spool of thread contained 250 cm of thread. A roll of thin wire contained 0.4 m of wire.

a. Convert both lengths to mm. Is there more thread or more wire? How much more?

b. How many rolls of wire will have the same total length as 8 spools of thread?

3-5 *Converting Smaller Units of Length to Larger Units*

STUDY GUIDE In Section 3-4 larger units of linear measure were converted to smaller units by multiplying by powers of 10 ($10^1 = 10$, $10^2 = 100$, or $10^3 = 1000$). It is also possible to convert *smaller* units to *larger* units by *dividing* by powers of 10. Division by powers of 10 can be done quickly by moving the decimal point to the *left* as many places as indicated by the exponent.

Recall that 1 mm = 0.1 cm, 1 cm = 0.01 m, and 1 mm = 0.001 m.
Millimeters can be changed to centimeters by *dividing by 10*.
Centimeters can be changed to meters by *dividing by 100*.
Millimeters can be changed to meters by *dividing by 1000*.

FACT FINDING

1. Smaller units of linear measure can be changed to larger units by

   _____ by powers of 10.

2. Millimeters can be changed to centimeters by dividing by _____ .

3. Centimeters can be changed to meters by _____ by 100.

4. Millimeters can be changed to _____ by dividing by 1000.

MODEL PROBLEMS

1. Convert 350 cm to meters.

   Solution: Since centimeters can be changed to meters by dividing by 100,

   $$\frac{350}{100} = 3.5 \text{ m.}$$

   Note that 350 can be thought of as 350., with a decimal point at the right end of the numeral. Division by 100 is done quickly by moving the decimal point *two* places to the *left*, since $100 = 10^2$. Therefore

   $$350. \div 100 = 3.50 \text{ or } 3.5.$$

2. Convert 273 mm to centimeters.

   Solution: Since millimeters can be changed to centimeters by dividing by 10,

   $$\frac{273}{10} = 27.3 \text{ cm}.$$

   The decimal point in 273. has been moved *one* place to the left.

EXERCISES

1. Convert each of the following measurements in mm to cm:

   a. 25 mm   **c.** 8 mm   **e.** 125.4 mm   **g.** 3500 mm   **i.** 56 mm

   b. 314 mm   **d.** 53.7 mm   **f.** 9.6 mm   **h.** 32.85 mm   **j.** 0.1 mm

2. Convert each of the following measurements in cm to m:

   a. 48 cm   **c.** 5500 cm   **e.** 72.6 cm   **g.** 8375 cm   **i.** 800 cm

   b. 345 cm   **d.** 6 cm   **f.** 576.25 cm   **h.** 5.3 cm   **j.** 100 cm

3. Convert each of the following measurements in mm to m:

   a. 1000 mm   **c.** 7650 mm   **e.** 95 mm   **g.** 3 mm   **i.** 10,000 mm

   b. 2450 mm   **d.** 635 mm   **f.** 10 mm   **h.** 234.5 mm   **j.** 500 mm

4. Convert each of the following measurements to the unit indicated:

   a. 76 mm = __ cm   **d.** 432 mm = __ cm   **g.** 2.8 mm = __ cm

   b. 830 cm = __ m   **e.** 3750 cm = __ m   **h.** 27.2 cm = __ m

   c. 1250 mm = __ m   **f.** 9875 mm = __ m   **i.** 640 mm = __ m

5. A dime has a thickness of approximately 1 mm. Express this measurement in cm.

6. A quarter has a thickness of approximately 2 mm. A roll of quarters contains 40 quarters. Calculate the height of a roll of quarters in mm and convert this result to cm.

7. Place the correct symbol ($<$, $=$, or $>$) between each pair of measurements:

   a. 85 mm __ 8.5 cm   **f.** 3000 mm __ 3 m

   b. 450 cm __ 45 m   **g.** 5000 mm __ 50 cm

   c. 6550 mm __ 655 m   **h.** 750 cm __ 7500 m

   d. 5 mm __ 0.5 cm   **i.** 37 mm __ 0.37 m

   e. 29 cm __ 2900 m   **j.** 1000 mm __ 1 m

8. Measure the length of each of the following line segments in mm. Convert each result to cm.

a. _____

b. _____

c. _____

d. _____

e. _____

EXTRA
FOR
EXPERTS

Dimes are approximately 1 mm thick and are packaged in rolls of 50. Quarters are approximately 2 mm thick and are packaged in rolls of 40. Answer these questions:

a. Which is longer, a roll of dimes or a roll of quarters? How much longer? Convert this answer to cm.

b. What is the value of one roll of dimes? one roll of quarters?

c. A store owner came into a bank to obtain rolls of coins for making change in his business. He asked for two rolls of quarters and three rolls of dimes. How much money did he have to give to the bank teller (in dollars)? What was the total length of the five rolls of coins he received? Express your answer in mm and in cm.

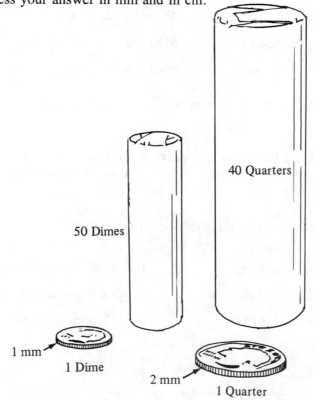

40 Quarters

50 Dimes

1 mm
1 Dime

2 mm
1 Quarter

3-6 *Comparing Meters and Kilometers*

The meter is too small a unit for measuring large distances, such as the distance between two cities. The unit commonly used for such measurements is the *kilometer,* which is equal to 1000 meters. The prefix "kilo-" means one thousand. This prefix is also used in measuring electrical power, in the terms "kilowatt" and "kilowatt-hour," for example.

A kilometer is a shorter distance than a mile. One kilometer is equal to approximately $\frac{5}{8}$ of a mile, or 8 kilometers are approximately equal to 5 miles.

Kilometers can be changed to meters by *multiplying by 1000*. Meters can be changed to kilometers by *dividing by 1000*. Study the following table:

100 m = 0.1 km
200 m = 0.2 km
300 m = 0.3 km
etc.
900 m = 0.9 km
1000 m = 1.0 km
2000 m = 2.0 km
etc.

FACT FINDING

1. The unit used for measuring large distances is the _____ .

2. One kilometer is equal to _____ meters.

3. The prefix "kilo-" means _____ .

4. One kilometer is _____ (<, =, or >) 1 mile.

5. Kilometers can be changed to meters by _____ by 1000.

6. Meters can be changed to kilometers by dividing by _____ .

MODEL PROBLEMS

1. Maria lives 3.2 km from the school. Express this distance in meters.

   Solution: Kilometers can be changed to meters by multiplying by 1000. Since $1000 = 10^3$, this multiplication can be done quickly by moving the decimal point three places to the right: $3.2 \times 1000 = 3.200. = 3200$ m. Note that it was necessary to place two zeros at the right end of the numeral 3.2 to do this multiplication.

2. A race in the Olympic Games is 5000 meters long. How many kilometers is this?

   Solution: Meters can be converted to kilometers by dividing by 1000. Since $1000 = 10^3$, this division can be done quickly by moving the decimal point three places to the left: $5000 = 5000.$ and $5000. \div 1000 = 5.000. = 5.000$ km or 5 km.

EXERCISES

1. Convert each of the following distances to meters:

   a. 7 km   **c.** 135 km   **e.** 27.3 km   **g.** 53.25 km   **i.** 0.65 km

   b. 12 km   **d.** 6.8 km   **f.** 2.75 km   **h.** 0.8 km   **j.** 0.375 km

2. Convert each of the following distances to kilometers:

   a. 4000 m   **c.** 250,000 m   **e.** 12,500 m   **g.** 1725 m   **i.** 7 m

   b. 37,000 m   **d.** 7500 m   **f.** 600 m   **h.** 50 m   **j.** 450 m

3. Patricia walks 1.75 km from her home to school each day. How many meters is this?

4. Frank drives 40 kilometers to work each day and the same distance home in the evening. Express the total distance he drives each day in meters.

5. Recall that 5 miles = 8 kilometers, approximately. The speed limit on American highways is 55 miles per hour. Express this number in kilometers per hour. $\left(Hint: \dfrac{5}{8} = \dfrac{55}{?}. \right)$

6. One of the races in the Olympic Games is 10,000 m in length. How many kilometers is this race?

7. The Empire State Building is one of the tallest buildings in the world. Its height is more than 380 meters. Express this height in kilometers.

8. From the Atlantic coast to the Pacific coast, the distance across the United States is about 4800 km. Express this distance in meters.

9. The distance around the world at the equator is approximately 40,000,000 meters. Express this distance in kilometers.

10. How many meters is $\frac{1}{2}$ kilometer? $\frac{1}{4}$ kilometer? $\frac{3}{4}$ kilometer?

EXTRA FOR EXPERTS

In a relay race, four swimmers had to swim 0.8 km each.

a. What is the total distance that the four swimmers swam?

b. How much of the race remained to be swum after the first swimmer had completed her swim and the second swimmer was half way through her swim?

c. After the first swimmer had swum 350 meters, how much further did she have to swim?

d. How much of the race had been completed when the third swimmer finished her swim? Express the answer:

    (1) in meters   (2) in kilometers   (3) as a fraction of the entire race

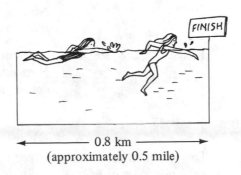

0.8 km
(approximately 0.5 mile)

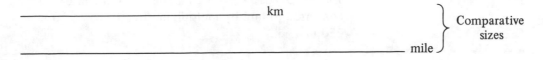

km

mile

Comparative sizes

3-7 Converting One Metric Unit of Length into Another

STUDY GUIDE In solving problems, it is often necessary to change from one unit of length to another. Changes may be made from millimeters to centimeters or to meters, from centimeters to meters, from meters to kilometers, from kilometers to meters, from meters to centimeters or to millimeters, and from centimeters to millimeters. All of these conversions are made by *multiplying* or *dividing by powers of 10*. Making such conversions frequently simplifies calculations in problem solving.

FACT FINDING

1. In problem solving, it is often necessary to change from one unit of _____ to another.

2. Changes may be made from millimeters to _____ or to _____ .

3. Changes may be made from centimeters to _____ or to _____ .

4. Changes may be made from meters to _____ or to _____ or to _____ .

5. Changes may be made from kilometers to _____ .

6. All of these changes are made by multiplying or _____ by _____ _____ _____ .

MODEL PROBLEMS

1. A piece of sheet metal is 1 meter long. It is necessary to cut it into four equal pieces to make a square frame. Express the length of each piece in (a) millimeters (b) centimeters (c) meters.

Solution: (a) 1 meter = 1 × 1000 mm = 1000 mm.

$$\frac{1000}{4} = 250 \text{ mm}$$

Note *multiplication* by a *power of 10*: $1000 = 10^3$.

(b) 1 meter = 1 × 100 cm = 100 cm

$$\frac{100}{4} = 25 \text{ cm}$$

Note *multiplication* by a *power of 10*: $100 = 10^2$.

(c) $\frac{1}{4} = 0.25$ m

This problem can also be solved in other ways. For example, do part (c) first to obtain 0.25 m. This length can be converted to cm by multiplying by 100: 0.25 × 100 = 25 cm. Then 25 cm can be converted to mm by multiplying by 10: 25 × 10 = 250 mm.

2. Convert 375 meters to (a) kilometers   (b) centimeters   (c) millimeters.

Solution: (a)   Meters can be converted to kilometers by *dividing* by 1000: $\frac{375}{1000} = 0.375$ km   Note *division* by a *power of 10:* $1000 = 10^3$.

(b)   Meters can be converted to centimeters by *multiplying* by 100: $375 \times 100 = 37,500$ cm.

(c)   Meters can be converted to millimeters by *multiplying* by 1000: $375 \times 1000 = 375,000$ mm.

EXERCISES

1. Convert each of the following lengths to the unit indicated:

   a. 8.35 km = ___ m   **f.** 68.3 m = ___ cm   **k.** 15 cm = ___ m

   b. 453 cm = ___ mm   **g.** 53 m = ___ mm   **l.** 850 m = ___ km

   c. 684 cm = ___ m   **h.** 2375 mm = ___ cm   **m.** 7.3 m = ___ cm

   d. 3750 m = ___ km   **i.** 16.3 km = ___ m   **n.** 5 m = ___ mm

   e. 4350 mm = ___ m   **j.** 38 cm = ___ mm   **o.** 57 mm = ___ cm

2. A board is 12 meters long. It is to be cut into eight equal pieces to make shelves. Express the length of each piece in:

   a. meters   **b.** centimeters   **c.** millimeters

3. Six pieces of pipe were purchased to do plumbing work. Each piece was 135 cm long. Express the total length of the pipe in:

   a. cm   **b.** m

4. A bookcase is to be built from two pieces of lumber, each 1.5 m long, and three pieces of lumber, each 1.25 m long. What is the total length of the lumber needed? Convert the answer from meters to centimeters.

5. Wooden molding is to be placed along the floor in a room, where the walls meet the floor. The room is 3.5 meters wide and 4.75 meters long. How many meters of molding should be purchased to complete this job? Convert the answer from meters to centimeters.

6. A sheet metal worker needed strips of metal of the following lengths for a special order: 87 cm, 2.75 m, 54 cm, 1.05 m, and 65 cm. Find the total length of metal needed.

7. Josephine ran three races in one day, of the following distances: 2.5 km, 5 km, and 10 km. What is the total number of *meters* that she ran in all three races?

8. A carpenter cut a piece of lumber of length 225 cm from a board 3.5 m in length. What length of board was left over?

9. A pipefitter had several pipes, each of length 65 cm. He needed to fit them together to run a pipe across a room 7.8 meters in length. How many pieces of pipe did he need?

10. A piece of chalk 105 mm in length is to be broken into three smaller pieces, all equal in length. How long will each of these smaller pieces be?

EXTRA
FOR
EXPERTS

The five players on a basketball team had the following heights: 189 cm, 194 cm, 2 m, 2.04 m, and 2.1 m. Answer these questions:

a. What is the total height of the five players?

b. How much taller is the tallest player than the shortest player?

c. Which two players are closest in height?

d. What is the average height of the five players?

3-8 *Units of Weight in the Metric System*

STUDY GUIDE The basic unit of weight in the metric system is the *gram*, which is a small unit. A gram is much smaller than an ounce, which is a basic unit of weight in the English system. One ounce is the equivalent of approximately 28.35 grams. An ordinary paper clip weighs about 1 gram.

In the English system, 1 pound = 16 ounces. Since 1 ounce = 28.35 grams, approximately, 16 ounces = 16 × 28.35 = 453.6 grams. Therefore 1 pound = 453.6 grams, approximately.

Since a gram is a small unit of weight, a larger unit is needed for use in the marketplace and elsewhere. This larger unit is the *kilogram*, and, as with units of length, the prefix ''kilo-'' means 1000. One kilogram = 1000 grams. One kilogram = 2.2 pounds, approximately.

Although the gram is a small unit of weight, sometimes even smaller units are needed, as, for example, to measure chemicals and ingredients in vitamins or medicines. A smaller unit than the gram is the *milligram*, and, as with units of length, the prefix "milli-" means $\frac{1}{1000}$ or 0.001. One gram = 1000 milligrams. One milligram = $\frac{1}{1000}$ gram or 0.001 gram.

Study the following table:

Unit	Abbreviation	Weight
kilogram	kg	1000 grams
gram	g	1 gram
milligram	mg	0.001 gram

Note that kilograms are obtained from grams by *dividing by 1000*, just as kilometers are obtained from meters by dividing by 1000.

Milligrams are obtained from grams by *multiplying by 1000*, just as millimeters are obtained from meters by multiplying by 1000.

The abbreviations for kilogram, gram, and milligram are not followed by periods.

Postage Stamp
weighed in
milligrams

First–Class Letter
1 ounce
(approximately 28.35 grams)

Mail Sack
About 32 kilograms
(approximately 70 pounds)

1. The basic unit of weight in the metric system is the _____.

2. A gram is _____ (<, =, or >) 1 ounce.

3. One pound _____ (<, =, or >) 500 grams.

4. One kilogram = _____ grams.

5. One kilogram _____ (<, =, or >) 1 pound.

6. One milligram = _____ gram.

7. One gram = _____ milligrams.

8. Kilograms are obtained from grams by dividing by _____.

9. Milligrams are obtained from grams by _____ by 1000.

MODEL PROBLEMS

1. A bag of candy weighed 400 grams. A bag of nuts weighed 1 pound. Which is the greater weight?

   Solution: A pound is more than 450 grams. The candy weighed only 400 grams. Therefore the bag of nuts weighed more than the bag of candy.

2. Indicate the appropriate unit for measuring the weight of each of the following, selecting from mg, g, and kg:

   a. Weight of an elephant   d. Weight of a student

   b. Weight of a pencil   e. Weight of a chemical in a vitamin tablet

   c. Weight of a dime

   Solution: a. kg   b. g   c. g   d. kg   e. mg

EXERCISES

1. The following words were in *italics* in this section. Look up their definitions in a dictionary and write their meanings:

   a. gram   b. kilogram   c. milligram

2. Place the symbol "<" or ">" between each pair of units to make the sentence true:

    a. milligram — gram   **d.** milligram — kilogram

    b. kilogram — gram   **e.** gram — milligram

    c. kilogram — milligram   **f.** gram — kilogram

3. A can of peas was marked "Net weight: 500 grams." Did the can contain more or less than 1 pound of peas?

4. Arrange these units of weight in order, from smallest to largest: g, kg, mg.

5. Next to each of the following write the abbreviation for the unit that would probably be used for measuring its weight. Choose from mg, g, kg.

    a. Weight of a large bag of apples   **i.** Weight of a baseball

    b. Weight of a box of cereal   **j.** Weight of a penny

    c. Weight of a fly   **k.** Weight of a dollar bill

    d. Weight of a notebook   **l.** Weight of a peanut

    e. Weight of a piece of chalk   **m.** Weight of a puppy

    f. Weight of a newborn baby   **n.** Weight of a tiger

    g. Weight of a ball-point pen   **o.** Weight of a piece of tissue paper

    h. Weight of a grain of salt

6. TRUE or FALSE?

    a. One ounce is less than 1 gram.   **c.** One kilogram is more than 2 pounds.

    b. One pound is more than 1 kilogram.   **d.** One ounce is less than 30 grams.

EXTRA FOR EXPERTS

A customer in a supermarket bought three cans of food, weighing 300 grams, 700 grams, and 1 kilogram. Another customer bought a bag of potatoes weighing 5 pounds. Which customer had the heavier weight to carry home?

3-9 *Comparing Grams and Kilograms*

STUDY GUIDE Since the gram is such a small unit of weight, the unit used most commonly in the marketplace is the *kilogram*, which is equal to 1000 grams. The prefix "kilo-" means one thousand. This prefix was also used in measuring length (kilometer).

A kilogram is a greater weight than a pound. One kilogram is equal to approximately 2.2 pounds. Since 1 pound is approximately equal to 453.6 grams, 2 pounds = 2 × 453.6 = 907.2 grams.

One kilogram = 1000 grams; therefore a kilogram is a little more than 2 pounds—in fact, about 2.2 pounds.

Kilograms can be changed to grams by *multiplying by 1000*.

Grams can be changed to kilograms by *dividing by 1000*.

Study the following table:

100 g	= 0.1 kg
200 g	= 0.2 kg
300 g	= 0.3 kg
etc.	
900 g	= 0.9 kg
1000 g	= 1.0 kg
2000 g	= 2.0 kg
etc.	

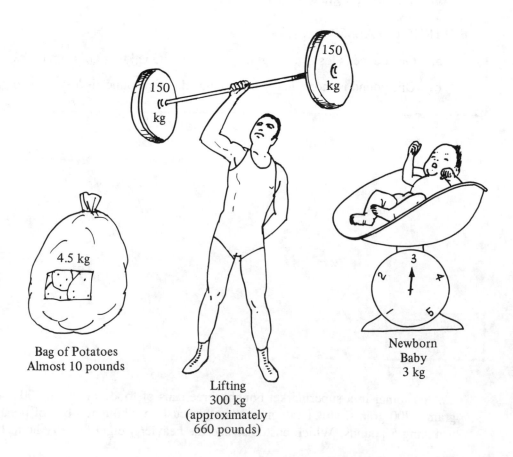

Bag of Potatoes
Almost 10 pounds

Lifting
300 kg
(approximately
660 pounds)

Newborn
Baby
3 kg

FACT FINDING

1. The unit of weight used most commonly in the marketplace is the

   _____ .

2. The prefix "kilo-" means _____ .

3. A kilogram is _____ ($<$, $=$, or $>$) a pound.

4. One kilogram is approximately equal to _____ pounds.

5. One kilogram = _____ grams.

6. Kilograms can be converted to grams by multiplying by _____ .

7. Grams can be converted to kilograms by _____ by 1000.

MODEL PROBLEMS

1. One bag of groceries weighed 1.35 kg, and another bag weighed 1300 g. Which bag was heavier?

   Solution: 1.35 kg = 1.35 × 1000 g = 1.350. or 1350 g.

   This bag was heavier than the one that weighed 1300 g. Note that kilograms were converted to grams by multiplication by 1000 = 10^3. For this reason, the decimal point in 1.35 was moved three places to the right, and it was necessary to provide a zero at the right end of the number.

   Alternative solution: 1300 g = 1300. ÷ 1000 kg = 1.300. or 1.3 kg, which is less than 1.35 kg, so the bag weighing 1300 g is lighter. In this method of solution, grams were converted to kilograms by division by 1000 = 10^3, so the decimal point was moved three places to the left in 1300.

2. A customer in a supermarket bought two bags of sugar, each weighing 1 kilogram. Another customer bought a 5-pound bag of sugar. Which customer bought more sugar?

   Solution: 1 kg = 2.2 pounds, approximately.
   2 kg = 2.2 × 2 pounds or 4.4 pounds, approximately.
   The customer who bought the 5-pound bag bought more sugar, since 5 > 4.4.

1. Convert each of the following weights to grams:

   a. 9 kg   **c.** 250 kg   **e.** 47.9 kg   **g.** 72.75 kg   **i.** 0.35 kg

   b. 32 kg   **d.** 3.2 kg   **f.** 6.25 kg   **h.** 0.6 kg   **j.** 0.875 kg

2. Convert each of the following weights to kilograms:

   a. 7000 g   **c.** 175,000 g   **e.** 37,500 g   **g.** 2225 g   **i.** 6 g

   b. 42,000 g   **d.** 3500 g   **f.** 300 g   **h.** 60 g   **j.** 675 g

3. A student weighed 58.2 kg. How many grams is this?

4. A student on the football team weighed 100 kg. Is this weight greater or less than 200 pounds? Express his weight in grams.

5. A package weighed 2350 g. Express this weight in kilograms.

6. A champion weightlifter can lift 300 kg. Express this weight in grams.

7. A truckdriver had to lift three packages onto his truck. Their weights were 18.3 kg, 12.95 kg, and 21 kg. Express the total weight he had to lift in kilograms and also in grams.

8. How many grams is $\frac{1}{2}$ kilogram? $\frac{1}{4}$ kilogram?

**EXTRA
FOR
EXPERTS**

Two grocery bags contained the following items:

Bag 1	Bag 2
1 box of rice—350 g	1 bag of apples—1.75 kg
1 bag of flour—2 kg	1 can of fruit cocktail—450 g
1 box of cereal—400 g	1 package of chicken—1250 g
1 bag of potatoes—4.5 kg	1 box of candy—2.5 kg

Which bag was heavier? How much heavier was it than the other bag?

3-10 *Comparing Grams and Milligrams*

STUDY GUIDE Although the gram is a small unit of weight, much smaller weights are needed in such fields as chemistry, medicine, and other sciences to measure very small quantities of chemicals or other substances. The unit commonly used to measure very small weights is the *milligram*, which is equal to $\frac{1}{1000}$ g or 0.001 g. One gram = 1000 milligrams. The prefix "milli-" means one thousandth. This prefix is also used in measuring length (<u>milli</u>meter).

Milligrams can be changed to grams by *dividing by 1000*.

Grams can be changed to milligrams by *multiplying by 1000*.

Study the following table:

$$100 \text{ mg} = 0.1 \text{ g}$$
$$200 \text{ mg} = 0.2 \text{ g}$$
$$300 \text{ mg} = 0.3 \text{ g}$$
$$\text{etc.}$$
$$900 \text{ mg} = 0.9 \text{ g}$$
$$1000 \text{ mg} = 1.0 \text{ g}$$
$$2000 \text{ mg} = 2.0 \text{ g}$$
$$\text{etc.}$$

Paper Clip
1 gram

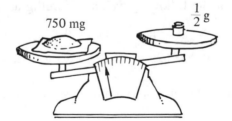

750 mg

$\frac{1}{2}$ g

500 mg

Vitamin
Tablet

FACT FINDING

1. The unit of weight used in measuring small quantities of chemicals is the

_____ .

2. The prefix "milli-" means _____ .

3. One gram = _____ milligrams.

4. Milligrams can be converted to grams by _____ by 1000.

5. Grams can be converted to milligrams by multiplying by _____ .

MODEL PROBLEMS

1. A chemical compound contained 350 mg of a certain chemical. What part of a gram is this?

   Solution: Milligrams can be converted to grams by dividing by 1000.

   $$350 \text{ mg} = 350. \div 1000 \text{ g} = 0.350. \text{ or } 0.35 \text{ g}$$

2. A scientist weighed 0.3 g of a particular chemical for an experiment. How many milligrams is this?

   Solution: Grams can be converted to milligrams by multiplying by 1000.

   $$0.3 \text{ g} = 0.3 \times 1000 \text{ mg} = 0.300. \text{ or } 300 \text{ mg}$$

EXERCISES

1. Convert each of the following weights to grams:

   a. 4000 mg   c. 8750 mg   e. 65 mg   g. 7 mg   i. 20,000 mg

   b. 6500 mg   d. 475 mg   f. 20 mg   h. 371.5 mg   j. 500 mg

2. Convert each of the following weights to milligrams:

   a. 7 g   c. 125 g   e. 0.4 g   g. 3.14 g   i. 18.9 g

   b. 48 g   d. 0.625 g   f. 22.3 g   h. 2.1 g   j. 25.98 g

3. A tablet contains 250 mg of vitamin C. What part of a gram is this?

4. A first-class letter weighed 30 grams. Is this weight more or less than 1 ounce? Express this weight in milligrams.

5. A certain vitamin tablet contains 200 mg of a particular chemical. What part of a gram is this? How many milligrams of the chemical are contained in three of the tablets? How many tablets must be taken by a person to consume 1 g of the chemical? 2 g?

6. One multivitamin tablet contains 500 mg of a certain vitamin. How many grams of this vitamin does a person consume if she takes four tablets?

7. A paper clip weighs about 1 g. How many milligrams does it weigh? How many milligrams would a box of 100 paper clips weigh?

8. For a certain experiment, 5.3 g of a chemical was needed. How many milligrams is this?

EXTRA FOR EXPERTS

One side of a scale (balance beam) contained 750 mg of a certain chemical. The other side of the scale contained a weight of $\frac{1}{2}$ g. How much more weight must be added to this side of the scale to place the scale in balance? (See the illustration on page 143.)

Measuring Your Progress

1. Indicate the unit most likely to be used to measure each of the following. Select from these units: mm, cm, m, km, mg, g, kg.

    a. Height of the Empire State Building   **f.** Weight of vitamin C in a tablet

    b. Thickness of a dime   **g.** Weight of a student

    c. Distance from New York to Washington   **h.** Distance across the Pacific Ocean

    d. Length of a ball-point pen   **i.** Weight of a postage stamp

    e. Weight of a first-class letter   **j.** Length of a finger

2. Write the numeral for each of the following:

    a. 10^4 **c.** 10^{-1} **e.** Seventeen and three tenths

    b. 10^0 **d.** 10^{-3} **f.** Five and eight hundredths

3. Write each of the following in words:

    a. 7.13   **b.** 0.27   **c.** 817   **d.** 23,502

4. What is the value of the 6 in each of the following numerals?

    a. 615   **b.** 3.06   **c.** 67,543   **d.** 29.6

5. Express each of the following as a decimal:

    a. $\frac{9}{10}$ **b.** $\frac{23}{100}$ **c.** $\frac{3}{100}$ **d.** $\frac{83}{1000}$

6. Express each of the following as a common fraction:

    a. 0.7   **b.** 0.03   **c.** 0.617   **d.** 0.57

7. Study the metric ruler shown. Match the measurements in Column A with the points listed in Column B.

A		B	
(1)	3 cm	a.	*E*
(2)	5 mm	b.	*C*
(3)	25 mm	c.	*B*
(4)	1.5 cm	d.	*D*
(5)	40 mm	e.	*A*

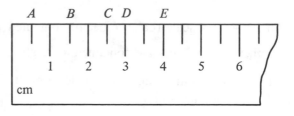

8. Convert each metric measure to its equivalent in the unit shown.

   a. 42 cm = __ mm
   b. 5.3 km = __ m
   c. 2000 mg = __ g
   d. 35.7 m = __ cm
   e. 25 kg = __ g
   f. 354 cm = __ m
   g. 3250 mm = __ m

   h. 454 g = __ kg
   i. 857 mm = __ cm
   j. 8.3 g = __ mg
   k. 235.8 cm = __ mm
   l. 40 km = __ m
   m. 1500 mg = __ g
   n. 8 m = __ cm

   o. 2.35 kg = __ g
   p. 28 cm = __ m
   q. 200 mm = __ m
   r. 73 g = __ kg
   s. 5 mm = __ cm
   t. 13 g = __ mg

9. Place the correct symbol (<, =, or >) between each pair of measurements.

   a. 0.9 cm __ 9 mm
   b. 2.2 kg __ 22,000 g
   c. 83.2 m __ 832 cm
   d. 50 mg __ 0.5 g
   e. 8.1 km __ 8100 m

   f. 500 mm __ 0.5 m
   g. 5000 g __ 50,000 mg
   h. 70 mm __ 700 cm
   i. 300 g __ 3 kg
   j. 75 cm __ 7.5 m

   k. 0.7 kg __ 700 g
   l. 1.3 km __ 1300 m
   m. 10 mm __ 1 cm
   n. 0.1 m __ 100 mm
   o. 2 g __ 2000 mg

10. The lengths of four pieces of lumber are as follows: 57 cm, 1.8 m, 92 cm, and 2.1 m. Express the total length of the four pieces in:

    a. meters
    b. centimeters

11. In a relay race, three runners ran distances of 500 meters each. Express the total distance run in:

    a. meters
    b. kilometers

12. In one hospital, five babies were born with the following weights: 2800 g, 3000 g, 3.5 kg, 3.8 kg, 4000 g. Express the total weight of the babies in:

    a. grams
    b. kilograms

13. A log 4 meters long was cut into eight pieces of equal length for firewood. Express the length of each piece in centimeters.

14. Five packages were mailed, each one weighing 1.3 kg. Express the total weight of the packages in grams.

15. Linda took one tablet of vitamin C each day for a week. Each tablet contained 250 mg of vitamin C. How much of this vitamin did Linda consume during the week?

16. A sewing needle was 45 mm in length. Express this length in:

   a. centimeters   b. meters

17. A first-class letter weighed 25 grams. Express the total weight of eight identical letters in:

   a. grams   b. kilograms

18. The weight of an ordinary paper clip is 1 gram. How much will one dozen paper clips weigh?

Measuring Your Vocabulary

Column II contains the meanings or descriptions of the words or phrases in Column I, which are used in this chapter.

For each number from Column I, write the letter from Column II that corresponds to the best meaning or description of the word or phrase.

Column I	Column II
1. Standard	a. Basic unit of weight in the metric system.
2. Metric system	b. Measure of length or distance.
3. Meter	c. $\frac{1}{1000}$ meter or 0.001 meter.
4. Linear measure	
5. Approximately	d. A symbol used to represent a number.
6. Kilometer	e. Numbers such as 10, 100, 1000.
7. Centimeter	f. Number system based on powers of 10.
8. Millimeter	g. The symbols 0, 1, 2, 3, 4, 5, 6, 7, 8, 9.
9. Power of 10	h. Established, acknowledged.
10. Decimal system	i. The value of a digit based on its location in a numeral.
11. Numeral	
12. Digit	j. Almost, nearly, about, close.
13. Place value	k. 1000 meters.
14. Exponent	l. A system used throughout the world for measuring lengths, distances, weights, and other values.
15. Gram	
16. Kilogram	
17. Milligram	m. $\frac{1}{1000}$ gram or 0.001 gram.
	n. $\frac{1}{100}$ meter or 0.01 meter.
	o. Basic unit of length in the metric system.
	p. A positive, zero, or negative numeral that may be used to express numbers as powers of 10.
	q. 1000 grams.

Chapter 4
Geometry

4-1 *Classifying Triangles by the Lengths of Their Sides*

STUDY GUIDE   A triangle is a three-sided, closed figure whose sides are straight line segments. Triangles are classified by the lengths of their sides and the sizes of their angles.

If no sides of the triangle are equal in length, the triangle is classified as a *scalene triangle*.

If the lengths of two of the sides of the triangle are equal, the triangle is classified as an *isosceles triangle*.

If the lengths of all three sides of the triangle are equal, the triangle is classified as an *equilateral triangle*.

In mathematics, we use figures and symbols to represent information. The symbol for triangle is △. A triangle is named by using *three capital letters*. The points where the sides of the triangle meet are called *vertices*. Each point is called a *vertex* of the triangle. Each vertex is named by using a capital letter.

Examples: In △ABC, A, B, and C are the vertices of the triangle. $\overline{AB}$, $\overline{BC}$, and $\overline{AC}$ are the sides of △ABC. In this example the sides are of equal length. To indicate that the line segments are equal in length, the same mark is used on each line segment. In each of the three isosceles triangles below, $\overline{DF} = \overline{FE}$.

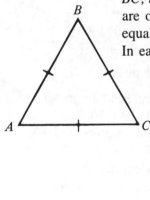

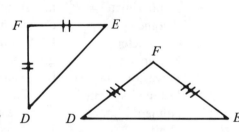

FACT FINDING

1.   A triangle is a _____ _____ figure whose sides are straight line segments.

2.   The symbol for a triangle is _____ .

3.   One way to classify triangles is by the _____ of their sides.

4.   If a triangle has _____ sides of equal length, it is an equilateral triangle.

5. If a triangle has _____ sides of equal length, it is a scalene triangle.

6. If a triangle has two sides of equal length, it is a(n) _____ triangle.

MODEL PROBLEM

Classify the triangles below according to the lengths of their sides.

a. 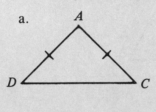   b. 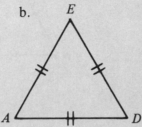   c.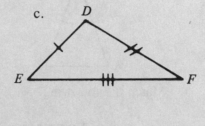

Solution: a. In $\triangle ADC$, $\overline{AD} = \overline{AC}$. Since two sides of $\triangle ADC$ are of equal length, $\triangle ADC$ is an isosceles triangle.

b. In $\triangle ADE$, $\overline{AD} = \overline{DE} = \overline{EA}$. Since all three sides of $\triangle ADE$ are of equal length, $\triangle ADE$ is an equilateral triangle.

c. In $\triangle DEF$, none of the sides are equal in length. Therefore $\triangle DEF$ is a scalene triangle.

‖

EXERCISES

1. Referring to the figure below, indicate whether each statement is TRUE or FALSE.

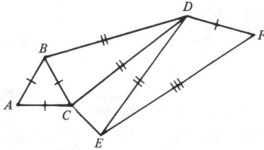

a. $\triangle ABC$ is an isosceles triangle.

b. $\triangle BDC$ is an isosceles triangle.

c. $\triangle CDE$ is an isosceles triangle.

d. $\triangle DFE$ is an isosceles triangle.

e. $\triangle ABC$ is an equilateral triangle.

f. $\triangle BCD$ is an equilateral triangle.

g. $\triangle CED$ is an equilateral triangle.

h. $\triangle DEF$ is an equilateral triangle.

i. $\triangle ABC$ is a scalene triangle.

j. $\triangle BDC$ is a scalene triangle.

k. $\triangle CDE$ is a scalene triangle.

l. $\triangle DEF$ is a scalene triangle.

2. State whether each triangle below is scalene, isosceles, or equilateral.

a.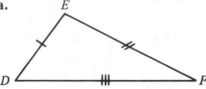

c.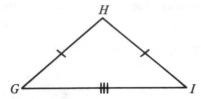

b.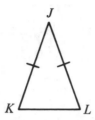

d.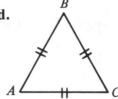

3. Match the following:

Column A	**Column B**
a.	Isosceles triangle
b. 	Scalene triangle
c. 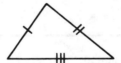	Equilateral triangle

4. Complete each of the following:

a. If △ABC has sides AB = 8″, BC = 6″, and CD = 13″, then

△ABC is a(n) _____ triangle.

b. In △DEF, if DE = 8″, $\overline{EF}$ must be _____ long in order for the triangle to be an equilateral triangle.

c. In △XYZ, if XY = 6 cm, YZ = 6 cm, and XZ = 9 cm, then △XYZ

is a(n) _____ triangle.

d.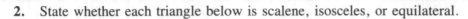

   (1) Name △I using three letters.

   (2) Name △II using three letters.

e. In △DEF, DE = 4.5″, EF = 7.1″, and DF = 7.1″. △DEF is

classified as a(n) _____ triangle.

f. In $\triangle XYZ$, $XY = 12$ cm, $YZ = 6$ cm, and $XZ = 12$ cm. $\triangle XYZ$ is classified as a(n) _____ triangle.

g. In $\triangle PQR$, $PQ = 3$ yards, $QR = 4$ yards, and $PR = 5$ yards. $\triangle PQR$ is classified as a(n) _____ triangle.

h. In $\triangle MNO$, $MN = 1$ mile, $\overline{NO} = \overline{MN}$, and $\overline{OM} = \overline{MN}$. $\triangle MNO$ is classified as a(n) _____ triangle.

i. In $\triangle RST$, $\overline{RS} = \overline{ST}$. Then $\triangle RST$ must be classified as a(n) _____ triangle. In order to classify the triangle as equilateral, what additional fact do you need to know?

5. $\triangle DEF$ is an equilateral triangle. List possible values for:

	a.	b.	c.
$DE =$	__	__	__
$EF =$	__	__	__
$FD =$	__	__	__

6.

	a.	b.	c.	d.	e.
AB	3″	6″	4 cm	6 yd.	8 m
BC	4″	6″	4 cm	8 yd.	8 m
CA	6″	6″	7 cm	10 yd.	5 m
Classify $\triangle ABC$					

1.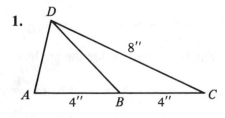

$\triangle ACD$ is a(n) _____ triangle since _____ _____.

2.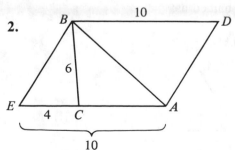

$\triangle ACB$ is a(n) _____ triangle since _____ _____.

3. Classify each triangle, using the information indicated on the figure:

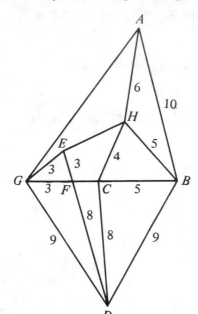

a. $\triangle ABH$

b. $\triangle CBD$

c. $\triangle BCH$

d. $\triangle GED$

e. $\triangle GEF$

f. $\triangle GFD$

4-2.1 *Naming Angles*

STUDY GUIDE An angle is a figure formed by two different rays having the same endpoint. This common endpoint is called the *vertex* of the angle, and the two rays are called the *sides* of the angle. The symbol "$\measuredangle$" or "$\angle$" designates the word "angle."

Example: In the angle shown below:

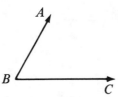

B is the vertex of the angle, and $\overrightarrow{BA}$ and $\overrightarrow{BC}$ are the sides of the angle.

An angle may be named in the following ways:

    (1) by its vertex,

    (2) by a lowercase letter or number placed inside the angle,

or   (3) by three capital letters.

When three capital letters are used, the middle letter *must* be the vertex letter.

Example: The angle may be named as:

    $\angle B$ or $\angle q$ or $\angle 3$

    or $\angle ABC$ or $\angle CBA$

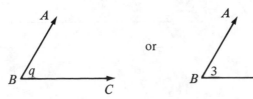

FACT FINDING

1. Indicate whether the statement is TRUE or FALSE.

    a. There is only one way to name an angle. _____

    b. An angle is formed by two distinct rays. _____

2. The common endpoint of the rays is called the _____ of the angle.

3. In this figure:

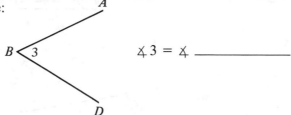

   $\angle 3 = \angle$ _____

4. In the figure above, name $\angle ABD$ in four ways.

    a. _____   b. _____   c. _____   d. _____

MODEL PROBLEM

Name the angles in the figure below.

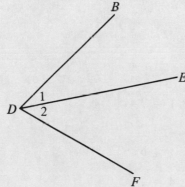

Solution: a. $\angle BDE = \angle EDB = \angle 1$.

b. $\angle 2 = \angle EDF = \angle FDE$.

c. $\angle BDF = \angle FDB$.

We would not name $\angle 1$ as $\angle D$ because there are at least three distinct angles at vertex D.

EXERCISES

1. Name the sides and the vertex of each of the following angles.

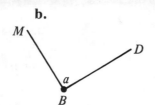

   a.   b.   c.   d.

2. Name each of the following angles.

a.

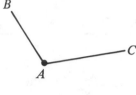

∢ _____

c.

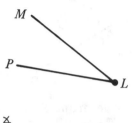

∢ _____

b.

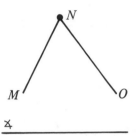

∢ _____

d.

∢ _____

3. Using the diagram at the left, use three letters to answer each question at the right.

a.

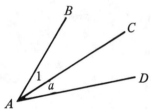

∢ 1 = ?

∢ a = ?

b.

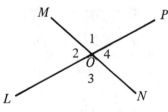

∢ 1 = ?

∢ 2 = ?

∢ 3 = ?

∢ 4 = ?

c.

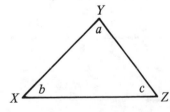

∢ a = ?

∢ b = ?

∢ c = ?

d.

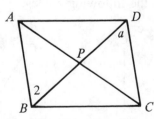

∢ 2 = ?

∢ a = ?

4-2.2 *Defining Angles by the Number of Degrees*

STUDY GUIDE   The *degree* is the unit of measure of angles. The symbol ''°'' is used to indicate degree. When a ray rotates from a position about its fixed endpoint until it reaches its original position, this complete rotation is defined as 360°. See the diagram below.

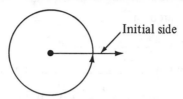

One degree is $\frac{1}{360}$ of a *rotation*. An angle may then be defined as the rotation of a fixed line (its initial side) about a fixed point (its vertex) until it reaches its final position (its terminal side).

Example: ∠*ABC* is an angle formed by rotating $\overrightarrow{BC}$ about point *B*

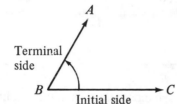

until it reaches $\overrightarrow{BA}$.
$\overrightarrow{BC}$ is its initial side.
$\overrightarrow{BA}$ is its terminal side.

Angles are classified according to their amount of rotation or according to the number of degrees contained in the angle.

A *right angle* is one fourth of a complete rotation. It is also an angle whose measure is 90°.

Example:

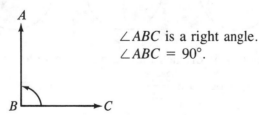

∠*ABC* is a right angle.
∠*ABC* = 90°.

An *acute angle* is less than a right angle.
Example:

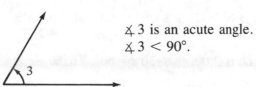

∡ 3 is an acute angle.
∡ 3 < 90°.

An *obtuse angle* is more than a right angle.
Example:

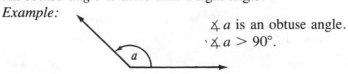

∡ *a* is an obtuse angle.
∡ *a* > 90°.

A *straight angle* is one half of a rotation. It is an angle whose measure is 180°.

Example: ∠B is a straight angle.
             ∠B = 180°.

B

FACT FINDING

1. Angles are measured in _____ .

2. The symbol for a _____ is °.

3. A complete rotation is _____ degrees.

4. A _____ angle is $\frac{1}{4}$ of a rotation.

5. One fourth of a rotation is _____ degrees.

6. An acute angle is _____ than a right angle.

7. An obtuse angle is _____ than a right angle.

8. A _____ angle is 180°.

MODEL PROBLEMS

Label the angles formed by the clock hands below as acute, right, obtuse, or straight.

a.   b.   c.   d.

Solution: (1) Since a clock is divided into 12 hours, there are 30° in each hour because $\frac{360°}{12} = 30°$.

(2) Estimate the time shown by the hands of the clock.

(3) Multiply the number of hours by 30°.

    a. Time is 3:00. 3 × 30° = 90°. The angle is a right angle.

    b. Time is 2:00. 2 × 30° = 60°. The angle is an acute angle.

    c. Time is 6:00. 6 × 30° = 180°. The angle is a straight angle.

    d. Time is 5:00. 5 × 30° = 150°. The angle is an obtuse angle.

1. Label as acute, right, obtuse, or straight, the angles formed by the hands of a clock when it is:

    a. 9 o'clock   **e.** 3:30   **h.** 10 o'clock

    b. 7 o'clock   **f.** 5 o'clock   **i.** 6 o'clock

    c. 2 o'clock   **g.** 12:05   **j.** 20 minutes to 12

    d. 3 o'clock

2. Classify the following angles as acute, obtuse, right, or straight:

    a. $\angle A = 56°$ **e.** $\angle E = 91°$ **h.** $\angle H = 89°$

    b. $\angle B = 180°$ **f.** $\angle F = 3°$ **i.** $\angle I = 110°$

    c. $\angle C = 75°$ **g.** $\angle G = 178°$ **j.** $\angle J = 137°$

    d. $\angle D = 90°$

3. Classify the angles in the following examples as acute, obtuse, right, or straight:

    a. The angle made by the wall and the ceiling.

    b. The angle of this car ramp:

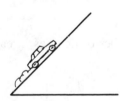

    c. The angle made by your pencil and this piece of paper when you are writing.

    d. The angle made by the blades of a fully open pair of scissors.

    e. The angle made by the pages of a newspaper open on a table.

4. Mr. Jackson said that his nephew was ''obtuse.'' Explain the meaning of the word ''obtuse'' in this sentence.

5. Harold Smither was rushed to St. John's Hospital because he had ''acute'' stomach pains. Explain the meaning of the word ''acute'' in this sentence.

MODEL PROBLEM

Using a protractor, find the number of degrees in $\angle A$.

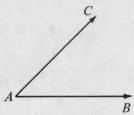

Solution:

(1) Place your protractor over the angle.

(2) Place the dot on the protractor over the vertex of the angle (A).

(3) Put the lower side of the protractor along the initial side of the angle (AB).

(4) Read the number where the terminal side of the angle intersects the edge of the protractor.

The number at the point of intersection is 45. Therefore $\angle A = 45$.

EXERCISES

1. Using a protractor, find the number of degrees in each of the following angles:

a.

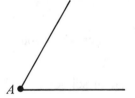

d.

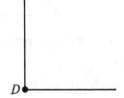

b.

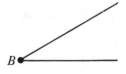

e.

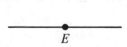

c.

f.

2. Use your protractor to make angles of these sizes:

a. 44° c. 106° e. 100° g. 78°

b. 83° d. 152° f. 121° h. 178°

3. If a pizza pie is divided equally among four friends, what is the size of the angle formed by each piece? See angle 1 in the diagram below.

4. If the angles between the spokes of the wheel shown are all equal, find the measure of ∠2.

5. Complete the chart below:

	Type of Angle	Number of Degrees
a.		Between 0° and 90°
b.	Right angle	
c.		110°
d.	Straight angle	

6. Find the number of degrees in the angle formed, and identify the angle as acute, right, obtuse, or straight, using the diagram below.

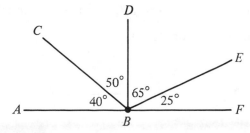

a. ∠ABC + ∠CBD = __   It is a(n) _____ angle.

b. ∠CBD + ∠DBE = __   It is a(n) _____ angle.

c. ∠CBD + ∠DBE + ∠EBF = __   It is a(n) _____ angle.

d. ∠DBF − ∠EBF = __   It is a(n) _____ angle.

7. ∠*ABC* is a straight angle in each of the following diagrams.

a.

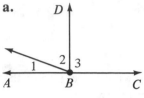

∠1 = 20°, ∠3 = 90°
Find the measure
of ∠2.

b.

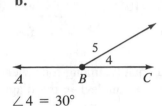

∠4 = 30°
Find the measure
of ∠5.

c.

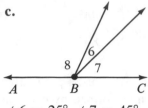

∠6 = 25°, ∠7 = 45°
Find the measure
of ∠8.

8. Complete the following statements, using your protractor. (The letter "m" stands for "measurement(s)."

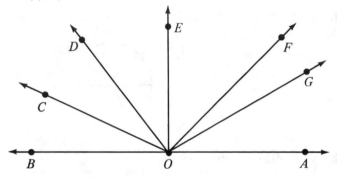

a. m∡*AOG* = __

b. m∡*BOD* = __

c. m∡*AOF* = __

d. m∡*EOF* = __

e. m∡*FOG* = __

f. m∡*AOB* = __

g. The m of ∡*EOF* plus the m of ∡*FOG* is the same as the m of

∡ __ or __°.

h. The m of ∡*BOD* minus the m of ∡*BOC* is the same as the m

of ∡ __ or __°.

i. The m of ∡*COF* minus the m of ∡*COD* is the same as the m

of ∡ __ or __°.

j. The sum of the m of ∡*COD*, ∡*DOE*, and ∡*EOF* is the same as the

m of ∡ __ or __°.

k. The m of ∡*AOF* plus the m of ∡*COF* minus the m of ∡*COD* is the

same as the m of ∡ __ or __°.

1. Answer the following questions:

   a. If a complete rotation of a ray is defined as 400°, how many degrees would there be in a right angle? How many degrees would there be in a straight angle?

   b. What is $\frac{1}{3}$ of a complete rotation of 400°?

   c. What is $\frac{1}{4}$ of 400°?

   d. What is $\frac{1}{9}$ of 400°?

   e. What is $\frac{1}{12}$ of 400°?

   f. Why do you think a complete rotation is traditionally defined as 360°?

2. Arrange these angles in size order with the smallest angle first:

   a. Straight angle   b. Right angle   c. Acute angle   d. Obtuse angle

3. Name five occupations that require a knowledge of how to measure angles.

4. Why does a carpenter use a miter box?

5. A skier has to choose among three trails to descend the mountain. She may choose advanced, intermediate, or beginner. Which of these trails would probably form the largest angle with the ground? Explain your answer.

4-3 *Sum of the Angles of a Triangle*

STUDY GUIDE   All triangles, regardless of their classification, have one property in common. The *sum* of the three angles of every triangle is equal to 180°:

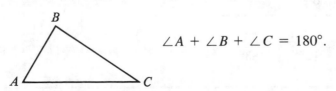

$$\angle A + \angle B + \angle C = 180°.$$

FACT FINDING

1. A triangle has _____ angles.

2. The sum of the three angles of a right triangle = _____°.

For Questions 3–6, answer TRUE or FALSE.

3. Every triangle has three equal angles. _____

4. The sum of the angles of an isosceles triangle = 180°. _____

5. If you know one angle of a triangle, you can find the other two angles. _____

6. If you know two angles of a triangle, you can find the third angle. _____

MODEL PROBLEM

Find the number of degrees in angle C.

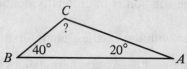

Solution: $\angle A + \angle B + \angle C = 180°$.

Since $\angle A = 20°$ and $\angle B = 40°$, then $20° + 40° + \angle C = 180°$.

$$60° + \angle C = 180°.$$

Answer: $\angle C = 120°$.

EXERCISES

1. Complete the chart. In $\triangle ABC$:

	$\angle A$	$\angle B$	$\angle C$
a.	70°	40°	
b.	65°		35°
c.	17°		96°
d.	42°	61°	
e.	76°	38°	
f.	45°		45°
g.	90°	60°	
h.	40°	90°	
i.	161°	6°	
j.		12°	83°

2. If a triangle has three equal angles, the measure of each angle is

_____ .

3. If a triangle has one right angle (see Exercise 1g and 1h) the other two

angles add up to _____°.

4. Find the measure of angle _t_ in each of the following:

a.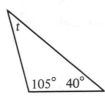

e.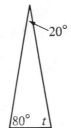

b.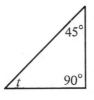

f.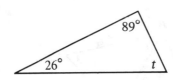

c.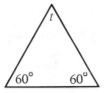

g.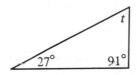

d.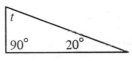

h.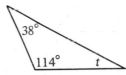

5. Indicate whether each of the following statements is TRUE or FALSE:

   a. A triangle may have two acute angles.

   b. A triangle may have three acute angles.

   c. A triangle may have two right angles.

   d. A triangle may have two obtuse angles.

6. What is the maximum number of acute angles a triangle can have? Why?

7. What is the maximum number of right angles a triangle can have? Why?

8. What is the maximum number of obtuse angles a triangle can have? Why?

9. What is the maximum number of straight angles a triangle can have?
Why?

10. Determine whether each set of angles can be the three angles of a triangle.

a. 40°, 30°, 70°

b. 66°, 74°, 40°

c. 100°, 10°, 30°

d. 40°, 30°, 100°

e. 90°, 90°

f. 45°, 45°, 45°

g. 100°, 90°, 10°

h. 87°, 13°, 80°

11. What is the third angle of a triangle if the first two angles are:

a. 80°, 10°

b. 41°, 47°

c. 57°, 36°

d. 45°, 40°

e. 90°, 45°

f. 67°, 84°

g. 18°, 111°

h. 96°, 54°

12.

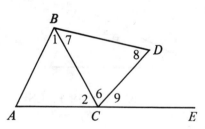

Given that ∠CAD is a straight angle:

a. ∡BAD = 110°.

∡1 = ___.

b. ∡B = 63°.

∡BCA = ___°.

c. ∡BCF = ___°.

d. ∡A = 65°.

∡1 = 55°.

∡2 = ___°.

e. ∡6 = 71°.

∡7 = 48°.

∡8 = ___°.

f. ∡9 = ___°.

g. ∡DBC is a right angle.

∡ABC is a straight angle.

∡ABD = ___°.

h. ∡C = 41°.

∡1 = ___°.

i. ∡A = 43°.

∡2 = ___°.

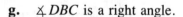

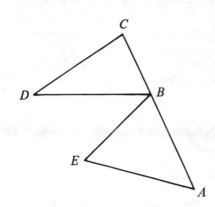

j. ∡*ABC* is a straight angle.

∡*CBA* = __°.

k. ∡*CDB* = 35°.

∡*CBD* = 65°.

∡*DCB* = __°.

l. ∡*DBE* = 45°.

∡*EBA* = __°.

m. ∡*BEA* = 60°.

∡*BAE* = __°.

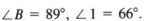

n. ∡*A* = 30°.

∡*C* is twice ∡*A*.

∡*B* = __°.

o. ∡*B* is three times ∡*A*.

∡*B* = __°.

EXTRA FOR EXPERTS

∠*B* = 89°, ∠1 = 66°.

1. Find the measure of ∠2.

2. If ∠*BCD* = 90°, find the measure of ∠3.

3. If ∠*CAD* = 24°, find the measure of ∠4.

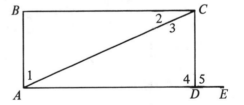

4. If ∠*ADE* is a straight angle, find the measure of ∠5.

5. In △*ABD*:
∡*BAD* = 90°.
∡*ABD* = 45°.

∡*ADB* = __°.

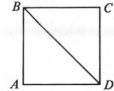

6. If ∡*ABC* is a right angle, ∡*CBD* = __°.

7. If ∡*CDA* is a right angle, ∡*CDB* = __°.

8. In △*BCD*: ∡*C* = __°.

9. What does the total measure of the angles of figure *ABCD* appear to be?

Find the value of x. Then, using the value you found for x, find the number of degrees in $\angle D$, $\angle E$ and $\angle A$.

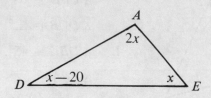

Solution: (1) $\angle A = 2x°$, $\angle D = (x - 20)°$, and $\angle E = x°$.
$\angle A + \angle D + \angle E = 180°$.

(2) Substitute: $2x° + (x - 20)° + x° = 180°$.

(3) Combine like terms: $4x - 20 = 180°$.

(4) Solve for x: $4x = 200$
$x = 50°$.

(5) Answer: $\angle A = 2(50) = 100°$.
$\angle D = (50 - 20) = 30°$.
$\angle E = 50°$.

EXTRA FOR EXPERTS

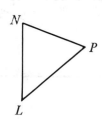

1. If $\angle L = x°$, $\angle P = (x + 10)°$, and $\angle N = (x + 20)°$, find the measures of $\angle L$, $\angle P$, and $\angle N$.

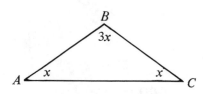

2. If $\angle A = x°$, $\angle C = x°$ and $\angle B = 3x°$, find the measures of $\angle A$, $\angle B$ and $\angle C$.

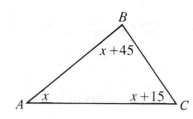

3. If $\angle A = x°$, $\angle B = (x + 45)°$, and $\angle C = (x + 15)°$, find the measures of $\angle A$, $\angle B$, and $\angle C$.

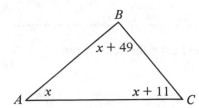

4. If $\angle A = x°$, $\angle B = (x + 49)°$, and $\angle C = (x + 11)°$, find the measures of $\angle A$, $\angle B$, and $\angle C$.

4-4.1 *Classifying Triangles by the Measure of Their Angles*

STUDY GUIDE We have classified triangles by the lengths of their sides. We can also classify them by the number of degrees in their angles.
  If a triangle has a right angle, it is called a *right* triangle.
If a triangle has an obtuse angle, it is called an *obtuse* triangle.
If a triangle has three acute angles, it is called an *acute* triangle.

FACT FINDING

1. If a triangle has two acute angles and one obtuse angle, it is a(n)

  _____ triangle.

2. If a triangle has three acute angles, it is a(n) _____ triangle.

3. If a triangle has two acute angles and one right angle, it is a(n)

  _____ triangle.

4. In $\triangle ABC$, $\angle A = 40°$, $\angle B = 67°$, and $\angle C = 73°$. $\triangle ABC$ may be

classified as _____ .

5. In $\triangle DEF$, $\angle D = 90°$; $\triangle DEF$ is a(n) _____ triangle.

6. A triangle may be classified according to its sides or its _____ .

MODEL PROBLEM

Classify each triangle according to the measure of its angles.

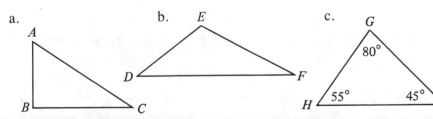

a.   b.   c.

Solution: a. $\angle B$ is a right angle, so $\triangle ABC$ is classified as a right triangle.

  b. $\angle E$ is an obtuse angle, so $\triangle DEF$ is classified as an obtuse triangle.

  c. $\angle G$, $\angle H$, and $\angle K$ are acute angles, so $\triangle GHK$ is classified as an acute triangle.

1. Classify each triangle according to its angles, and give the reason for your answer.

a.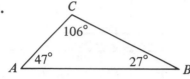

f.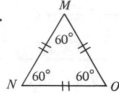

b.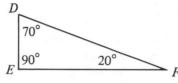

g.

c.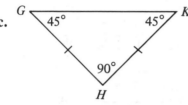

h.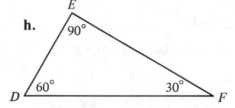

d.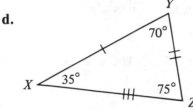

i.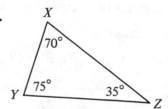

e.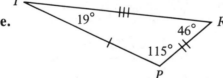

j.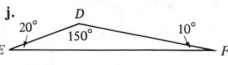

2. In △ABC, ∡A = 120°, ∡B = 30°, and ∡C = 30°. Then △ABC is _____. Why?

3. In △ABC, ∡A = 36°, ∡B = 72°, ∡C = __°. △ABC is _____. Why?

4. In △ABC, ∡A = 60°, ∡B = 60°, and ∡C = __°. △ABC is _____. Why?

5. In △DEF, ∡D = 90°, ∡E = 25°. △DEF is _____. Why?

6. Complete the following chart:

	m∡A	m∡B	m∡C	△ABC is
a.	70°	40°	70°	
b.		20°	70°	
c.	140°	10°	30°	
d.		65°	50°	
e.	91°	75°		

7. Draw an example of each of the following triangles. Using your protractor, verify that you have classified each one properly.

    a. Right triangle   **c.** Acute triangle

    b. Obtuse triangle   **d.** Equiangular triangle

8. YES or NO? May a triangle have:

    a. Two obtuse angles?   **g.** A right angle and two equal sides?

    b. Three right angles?

    c. Three acute angles?   **h.** An obtuse angle and two equal sides?

    d. One obtuse and one right angle?

    e. A right angle and no equal sides?   **i.** Only one acute angle?

    f. An obtuse angle and no equal sides?

9. Using your protractor to measure the angles, classify the following triangles:

    a.
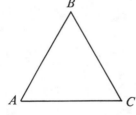

    d.
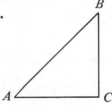

    b.
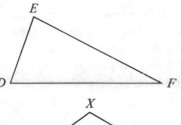

    e.

    c.

    f.
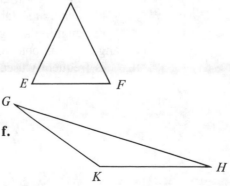

10. Referring to the diagrams, choose the number or numbers that correctly match each of the following:

a. A triangle with three acute angles

b. A triangle with an obtuse angle and two acute angles

c. A triangle with three equal angles

d. A triangle with two equal angles

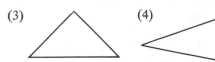

EXTRA FOR EXPERTS

1. In $\triangle ABC$, $\angle A = 90°$. If $\angle C = x°$, represent $\angle B$ in terms of $x°$.

2. In $\triangle ABC$, $\angle A = x°$, $\angle B = (2x - 10)°$, and $\angle C = (3x + 10)°$. Find the number of degrees in each angle, and classify the triangle.

3. In $\triangle ABC$, $\angle A = x°$, $\angle B = 3x°$, and $\angle C = 5x°$. Find the number of degrees in each angle, and classify the triangle.

4. Why is a building erected at a right angle to the ground?

4-4.2 *Right Triangles*

STUDY GUIDE In a right triangle, the side opposite the right angle is called the *hypotenuse* and the other two sides are the *legs*. In a right triangle, the right angle is the largest angle. The side opposite that angle, the hypotenuse, is the longest side of the triangle. The symbol "∟" is frequently used to represent a right angle.

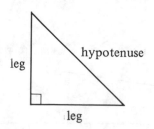

FACT FINDING

1. The side opposite the right angle in a right triangle is called the

_____ .

2. The right angle in a right triangle is formed by the _____ of
the triangle.

3. The longest side of a right triangle lies opposite the _____

and is called the _____ .

MODEL PROBLEMS

1. In right triangle ABC, if $\angle B$ is the right angle, identify side $\overline{AC}$.

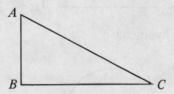

Solution: $\overline{AC}$ is the hypotenuse of right triangle ABC because it is
opposite angle B.

2. Using the letters R, S and T, label the triangle below so that $\overline{RS}$ is the
hypotenuse of the triangle.

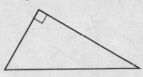

Solution:

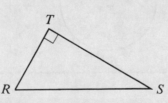

EXERCISES

1. Name the hypotenuse and legs of each right triangle below.

a.

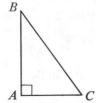

b.

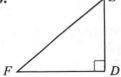

2. a. In rt. △*AXS*, name the hypotenuse.

b. In rt. △*AXR*, name the hypotenuse.

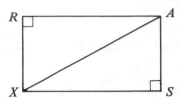

3. a. In rt. △*XYZ*, name the hypotenuse.

b. In rt. △*YZB*, name the hypotenuse.

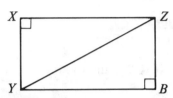

4. In rt. △*AOB*, name the hypotenuse.

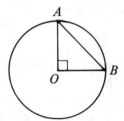

5. In rt. △*DEF*, name the hypotenuse.

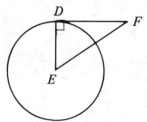

EXTRA FOR EXPERTS

1. In right △*XYZ*, if ∠*Z* is the right angle, ∠*X* must be a(n) _____ angle and ∠*Y* must be a(n) _____ angle.

2. In right △*PQR*, if $\overline{PQ}$ is the hypotenuse, name the right angle.

3. In right △*RST*, if $\overline{RS}$ is the longest side of the triangle, name the right angle.

4-5 *The Pythagorean Rule*

Pythagoras was a Greek mathematician who lived during the sixth century B.C. The rule for the relationship among the sides of a right triangle was named for him. The *Pythagorean rule* can be expressed as follows:

The sum of the squares of the lengths of the legs of a right triangle equals the square of the length of the hypotenuse.
This can be written as:

$$(\text{leg})^2 + (\text{leg})^2 = (\text{hypotenuse})^2.$$

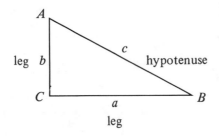

If leg $\overline{CB}$ is represented by a lower case a, leg $\overline{CA}$ is represented by a lower case b, and hypotenuse $\overline{AB}$ is represented by a lower case c, the rule may be written as

$$a^2 + b^2 = c^2.$$

FACT FINDING

1. The rule stating the relationship among the _____ of a right triangle was named for _____ .

2. The _____ of the right triangle are the two sides of the triangle opposite the two acute angles.

3. The _____ of the squares of the lengths of the legs of a right triangle equals the square of the length of the hypotenuse.

4. In a right triangle, $(\text{leg})^2 + (\text{leg})^2 =$ _____ .

MODEL PROBLEM

Find the square of 4.

Solution: $4^2 = 4 \times 4 = 16.$

Complete the following table:

	Number	Squared		Number	Squared
a.	1	$1^2 = 1 \times 1 = 1$	**p.**	20	
b.	2	$2^2 = 2 \times 2 = 4$	**q.**	25	
c.	3		**r.**	16	
d.	4		**s.**	100	
e.	5		**t.**	50	
f.	6		**u.**	x	
g.	7		**v.**	$\frac{1}{2}$	
h.	8		**w.**	$\frac{1}{3}$	
i.	9		**x.**	$\frac{2}{5}$	
j.	10		**y.**	$\frac{3}{7}$	
k.	11		**z.**	0.4	
l.	12		**aa.**	0.5	
m.	13		**bb.**	0.01	
n.	14		**cc.**	0.03	
o.	15		**dd.**	0.08	

MODEL PROBLEMS

A perfect square can be expressed as the product of two equal factors.

1. Is 4 a perfect square?

    Solution: $4 = 2 \times 2$ 4 is the product of two equal factors, 2 and 2. 4 is a perfect square.

2. Is 10 a perfect square?

    Solution: $10 = 5 \times 2$ 10 cannot be expressed as the product of two equal factors. 10 is *not* a perfect square.

Express each of the following as the product of two equal factors where possible:

a. $49 = 7 \times 7$ f. $1 =$ k. $104 =$ p. $10 =$

b. $36 =$ g. $4 =$ l. $0.09 =$ q. $15 =$

c. $64 =$ h. $\dfrac{1}{4} =$ m. $0.25 =$ r. $\dfrac{1}{100} =$

d. $100 =$ i. $\dfrac{4}{9} =$ n. $0.36 =$ s. $\dfrac{1}{25} =$

e. $25 =$ j. $0.01 =$ o. $0.3 =$

MODEL PROBLEM

In a triangle, if the longest side squared equals the sum of the squares of the other two sides, the triangle is a right triangle.

In $\triangle ABC$ below, if $c = 7$, $a = 4$ and $b = 5$, is $\triangle ABC$ a right triangle?

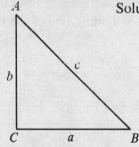

Solution: Does $c^2 = a^2 + b^2$ (since c is the longest side of the triangle)?
Does $7^2 = 4^2 + 5^2$?
Does $49 = 16 + 25$?
Does $49 = 41$?
Since 49 does *not* equal 41, $\triangle ABC$ is *not* a right triangle.

For each of the following, determine whether a triangle with sides of the lengths given is a right triangle. (*Remember:* In the rule $a^2 + b^2 = c^2$, the length of the longest side is c.)

a. 3, 4, 7   c. 5, 4, 6   e. 20, 21, 29

b. 5, 4, 3   d. 24, 11, 12

EXTRA FOR EXPERTS

1. If $c = 6$, find c^3.

2. If $c = \dfrac{1}{2}$, find c^3.

3. Can c^3 ever be a perfect square? Explain.

4. a. On a piece of graph paper, draw a right triangle with sides 3, 4, and 5. Use the markings on the paper to find sides 3, 4, and 5.

    b. Draw a square on each side of the triangle. Find the area of each of the squares that you have drawn.

    c. What is the relationship among the sizes of these squares?

4-6 *Using the Pythagorean Rule*

STUDY GUIDE The symbol "$\sqrt{}$" means *square root*. Finding the square root of a number means finding one of its two equal factors. The $\sqrt{9}$ is equal to 3 since $3 \times 3 = 9$.

FACT FINDING

1. The symbol "$\sqrt{}$" means _____ _____.

2. To find the square root of a number means finding one of the two equal

    _____ of the number.

MODEL PROBLEMS

1. Find $\sqrt{49}$.

    Solution: $49 = 7 \times 7$. $\sqrt{49} = \sqrt{7 \times 7} = 7$.

2. Find n if $\sqrt{n} = 6$.

    Solution: $36 = 6 \times 6$. $\sqrt{36} = \sqrt{6 \times 6} = 6$.

EXERCISES

1. Find:

a. $\sqrt{25}$	**f.** $\sqrt{625}$	**k.** $\sqrt{4}$	**p.** $\sqrt{0.01}$
b. $\sqrt{16}$	**g.** $\sqrt{100}$	**l.** $\sqrt{1}$	**q.** $\sqrt{0.25}$
c. $\sqrt{9}$	**h.** $\sqrt{64}$	**m.** $\sqrt{\frac{1}{25}}$	**r.** $\sqrt{0.09}$
d. $\sqrt{144}$	**i.** $\sqrt{121}$	**n.** $\sqrt{\frac{1}{4}}$	**s.** $\sqrt{0.16}$
e. $\sqrt{225}$	**j.** $\sqrt{36}$	**o.** $\sqrt{\frac{1}{16}}$	**t.** $\sqrt{0.0121}$

2. Find:

   a. If $n = 4$, $\sqrt{n} = \_\_$ **e.** If $n = 7$, $n^2 = \_\_$

   b. If $n = 9$, $\sqrt{n} = \_\_$ **f.** If $n = \dfrac{3}{4}$, $n^2 = \_\_$

   c. If $n = 36$, $\sqrt{n} = \_\_$ **g.** If $n = \dfrac{1}{16}$, $n^2 = \_\_$

   d. If $n = 3$, $\sqrt{n} = \_\_$

MODEL PROBLEM

Find the hypotenuse of a right triangle if one leg = 15 and the other = 20.

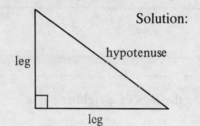

Solution: $(\text{leg})^2 + (\text{leg})^2 = (\text{hypotenuse})^2$

$$15^2 + 20^2 = c^2$$
$$225 + 400 = c^2$$
$$625 = c^2$$
$$\sqrt{625} = c$$
$$25 = c$$

EXERCISE

Find the hypotenuse of each right triangle whose legs are given below. Estimate the length of the hypotenuse, and verify your answer by multiplication.

	Leg$_1$	Leg$_2$	Hypotenuse
a.	3	4	
b.	5	12	
c.	6	8	
d.	10	24	
e.	8	15	
f.	15	36	
g.	12	16	
h.	16	30	
i.	14	48	

A 25-foot ladder is leaning against a wall. If the foot of the ladder is 7 feet from the wall, how far up the wall will the ladder reach?

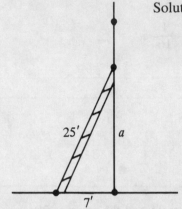

Solution: hypotenuse = 25, leg = 7
$$(Leg)^2 + (leg)^2 = (hypotenuse)^2$$
$$a^2 + 7^2 = 25^2$$
$$a^2 + 49 = 625$$
$$a^2 = 576$$
$$a = \sqrt{576}$$
$$a = 24'$$

EXERCISES

1. Find the missing values below if a and b are the legs of a right triangle and c is the hypotenuse of the triangle.

	a	b	c
a.		8	10
b.	10		26
c.		36	39
d.		12	15
e.	15		25
f.		48	50
g.		4	5
h.	8		10

2. Find the length of x in each right triangle.

a.

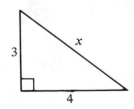

b.

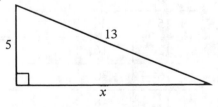

c.

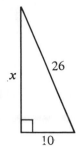

d.

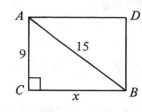

3. Solve the following problems:

   a. If the foot of a ladder is 16 feet from the wall, how high up on the wall does a 20-foot ladder reach?

   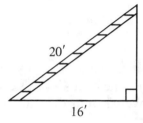

   b. In a marathon, a woman starting at *A* runs 5 miles east, then 12 miles due north. Another runner goes directly from point *A* to point *C*.

   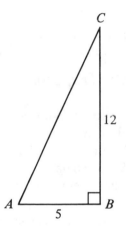

   (1) What distance does the first runner run?

   (2) What distance does the second runner run?

   (3) How many fewer miles does the second runner run?

   c. Will a nonfolding umbrella that is 23″ long fit into a carrying case that is 20″ by 15″? Explain your answer.

   d. Spider Man climbed 30 feet up the Sears Building, injured his arm, and requested that a ladder be placed against the building for him to descend. If the ladder is placed 16 feet from the foot of the building, how long a ladder must be used to assist him in descending?

1. An electric cable must be installed between points A and B. If $BC = 40$ m and $AC = 9$ m, how many meters of cable are needed to do the job?

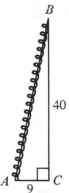

2. A motorboat travels across a river that is 250 yards wide. As it travels across the river, the current carries the boat 40 yards downstream. What is the distance, to the nearest yard, covered from the starting point?

3. A dune buggy traveled 9 miles west and then 6 miles south. How far was it, to the nearest mile, from the starting point?

4-7.1 *Quadrilaterals: Parallelogram, Trapezoid*

STUDY GUIDE

A *quadrilateral* is a four-sided, closed figure whose sides are line segments. A quadrilateral is named by listing its vertices consecutively. The quadrilateral at the left may be named $ABCD$, $BCDA$, $BADC$, $ADCB$, and so on. It may *not* be named $ACBD$ because in this arrangement the vertices are *not* listed consecutively.

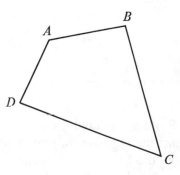

Many special quadrilaterals have sides that are parallel. *Parallel lines* lie in the same plane and do not meet. Two examples of parallel lines are the right and left edges of this page and the lines on notebook paper. If $\overline{AB}$ is parallel to $\overline{CD}$, it is indicated on the diagram by arrows:

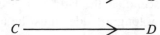

and represented as $\overline{AB} \parallel \overline{CD}$. The symbol "$\parallel$" is read as "is parallel to."

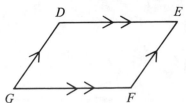

A quadrilateral whose opposite sides are parallel is called a *parallelogram*. Quadrilateral $DEFG$ at the left is a parallelogram because $\overline{DG} \parallel \overline{EF}$ and $\overline{DE} \parallel \overline{GF}$.

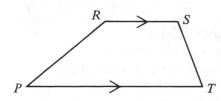

If only one pair of opposite sides of a quadrilateral is parallel, the figure is a *trapezoid*. In trapezoid *RSTP* at the left, $\overline{RS} \parallel \overline{PT}$, that is, $\overline{RS}$ is parallel to $\overline{PT}$. $\overline{RS}$ and $\overline{PT}$ are the bases of the trapezoid. The nonparallel sides, $\overline{RP}$ and $\overline{ST}$, are called legs. If the nonparallel sides of the trapezoid are equal, the trapezoid is called an *isosceles trapezoid*.

FACT FINDING

1. A parallelogram is a _____ sided figure.

2. In a _____ both pairs of opposite sides are parallel.

3. If one pair of opposite sides of a quadrilateral is parallel, the quadrilateral is called a _____ .

4. A trapezoid has _____ sides.

5. The symbol "$\parallel$" is used to represent _____ lines.

6. A quadrilateral is named by listing its _____ in consecutive order.

7. An isosceles trapezoid has two _____ sides and two _____ sides.

8. In a trapezoid the parallel sides are called _____ .

MODEL PROBLEM

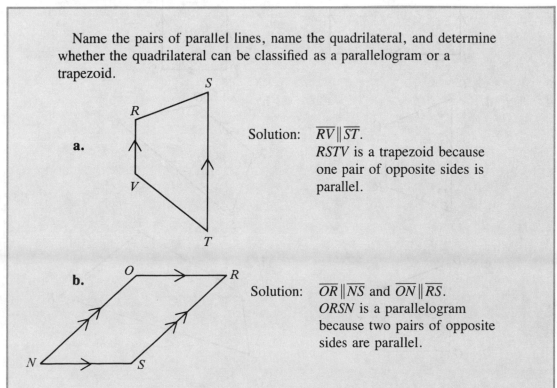

Name the pairs of parallel lines, name the quadrilateral, and determine whether the quadrilateral can be classified as a parallelogram or a trapezoid.

a.

Solution: $\overline{RV} \parallel \overline{ST}$.
RSTV is a trapezoid because one pair of opposite sides is parallel.

b.

Solution: $\overline{OR} \parallel \overline{NS}$ and $\overline{ON} \parallel \overline{RS}$.
ORSN is a parallelogram because two pairs of opposite sides are parallel.

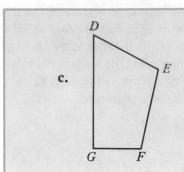

c.

Solution: No opposite sides are parallel. *DEFG* is a quadrilateral, and only a quadrilateral, because it has four line segments as sides and no two are marked parallel.

EXERCISES

1. Name the parallel lines, name the quadrilateral, and classify as a quadrilateral, a trapezoid, an isosceles trapezoid, or a parallelogram.

a.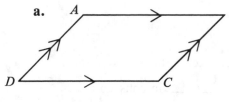

Solution: $\overline{AB} \parallel \overline{CD}$, $\overline{AD} \parallel \overline{BC}$. *ABCD* is a quadrilateral. Quad. *ABCD* is a parallelogram.

b.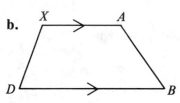

f.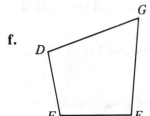

c.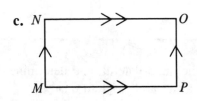

g.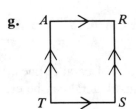

d.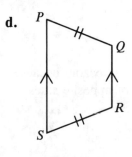

h.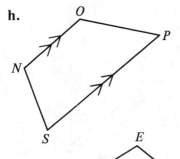

e.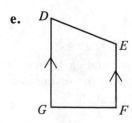

i.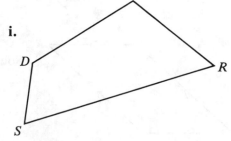

2. Draw quadrilateral *ABCD* so that:

    a. *ABCD* is a parallelogram.

    b. *ABCD* is a trapezoid.

    c. *ABCD* is an isosceles trapezoid.

    d. Name the pair of parallel sides in Exercise 2a, 2b, and 2c.

3. Draw trapezoid *DEFG*, label the trapezoid, and name the pair of parallel sides.

4. Draw parallelogram *XARS*, label the parallelogram, and name the pairs of parallel sides.

5. Draw isosceles trapezoid *NRST*, label the isosceles trapezoid, and name the bases and the legs.

6. Draw quadrilateral *EFGA*.

7. Draw quadrilateral *TVSR* and name the opposite vertices.

8. Draw parallelogram *DRAB*, label the parallelogram, and name the pairs of parallel sides.

9. Draw parallelogram *RNOS*, label the parallelogram, and name the pairs of parallel sides.

10. The quadrilateral, trapezoid, and parallelogram all have four sides. Explain how they are different.

11. Answer YES or NO:

    a. Is every quadrilateral a parallelogram?

    b. Is every parallelogram a quadrilateral?

    c. Is every trapezoid a parallelogram?

    d. Is every parallelogram a trapezoid?

4-7.2 *Special Parallelograms: Rhombus, Rectangle*

STUDY GUIDE Quadrilaterals with equal sides are classified more particularly. An isosceles trapezoid is a trapezoid with the pair of nonparallel opposite sides equal. A parallelogram has opposite sides equal as well as parallel.

A *rhombus* is a parallelogram with all sides equal. *ABCD* is a rhombus. $\overline{AB} = \overline{BC} = \overline{CD} = \overline{AD}$, $\overline{AB} \| \overline{CD}$, and $\overline{AD} \| \overline{BC}$. A quadrilateral with four equal sides is equilateral. A rhombus is equilateral.

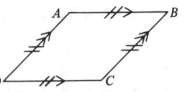

A parallelogram with four equal angles is a *rectangle*. *ARST* is a rectangle. $\overline{AR} \| \overline{ST}$, $\overline{AT} \| \overline{RS}$, and $\angle A = \angle R = \angle S = \angle T$. A quadrilateral with four equal angles is equiangular. A rectangle is equiangular.

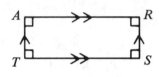

A *square* is a parallelogram with all sides and all angles equal. *FGHE* is a square. $\overline{FG} \| \overline{EH}$, $\overline{GH} \| \overline{FE}$, $\overline{EF} = \overline{FG} = \overline{GH} = \overline{EH}$, and $\angle E = \angle F = \angle G = \angle H$. A square is both equilateral and equiangular.

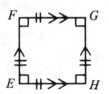

FACT FINDING

1. A parallelogram has two pairs of opposite sides that are _____ and _____ .

2. A _____ has four equal sides.

3. A rectangle has _____ equal angles.

4. A square has _____ equal sides and _____ equal angles.

5. An equilateral quadrilateral has _____ equal sides.

6. An equiangular quadrilateral has _____ equal angles.

7. A _____ is a parallelogram that is both equilateral and equiangular.

MODEL PROBLEM

For each figure, list all the names that the figure may have and underline the most descriptive name.

a.

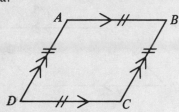

Solution: ABCD is a quadrilateral,
a parallelogram,
a rhombus.
ABCD is a rhombus because all four sides of the parallelogram are equal.
ABCD is a parallelogram because it is a quadrilateral with opposite sides parallel.
ABCD is a quadrilateral because it has four sides.

b.

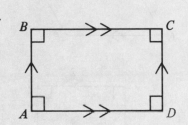

Solution: ABCD is a quadrilateral,
a parallelogram,
a rectangle.
ABCD is a rectangle because all four angles of the parallclogram are equal.

c.

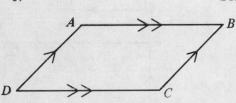

Solution: ABCD is a quadrilateral,
a parallelogram.
ABCD is a parallelogram because $\overline{AB} \parallel \overline{CD}$ and $\overline{AD} \parallel \overline{BC}$.

d.

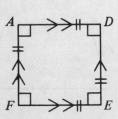

Solution: ADEF is a quadrilateral,
a parallelogram,
a rectangle,
a rhombus,
a square.
ADEF is a parallelogram because $\overline{AD} \parallel \overline{EF}$ and $\overline{AF} \parallel \overline{ED}$.
$\overline{EF} = \overline{DE} = \overline{AD} = \overline{AF}$ and $\angle A = \angle D = \angle E = \angle F$.
ADEF is a square because it is an equilateral and an equiangular parallelogram.

1. Identify the following figures, using as many of these terms as possible for each figure: quadrilateral, parallelogram, trapezoid, isosceles trapezoid, square, rectangle, rhombus. From the terms you list for each example, indicate which one of them is the most descriptive.

a.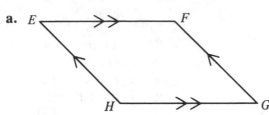

g.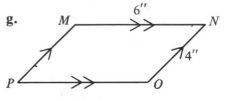

b.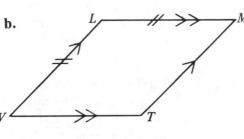

h.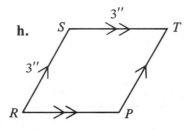

c.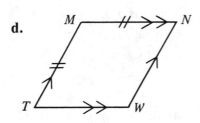

i.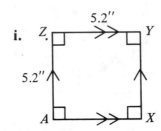

d.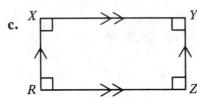

j.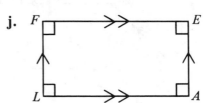

e.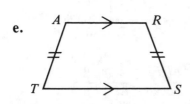

k.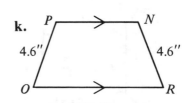

f.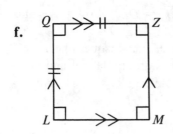

l.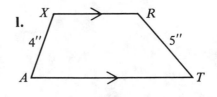

2. Find the length of the missing side.

    a. If *ABCD* is a square, *BC* = ___".

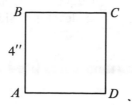

    b. If *ADEF* is a rectangle, *AF* = ___" and *FE* = ___".

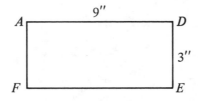

    c. If *RSTV* is a rhombus, *RS* = ___", *ST* = ___", and *TV* = ___".

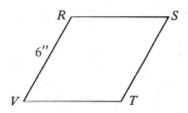

    d. If *ARST* is a square, ∠*R* = ___° and *AR* = ___".

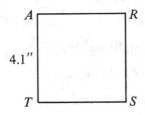

3. Answer YES or NO.

    a. Are all squares equilateral?

    b. Are all rectangles equilateral?

    c. Are all rhombuses equilateral?

    d. Are all rhombuses equiangular?

    e. Are all parallelograms equilateral?

    f. Are all parallelograms equiangular?

    g. Are all quadrilaterals equiangular?

4. List three words, other than "equilateral" and "equiangular," that begin with "equ," and write the definitions of these words.

5. Which properties are true for each figure named below?

	4 Sides	4 Equal Sides	Opposite Sides ∥	Opposite Sides =	4 Equal Angles
a. Quadrilateral					
b. Parallelogram					
c. Trapezoid					
d. Rhombus					
e. Rectangle					
f. Square					

 **EXTRA FOR EXPERTS**

1. Answer TRUE or FALSE.

    a. Every rhombus is a parallelogram.

    b. Every parallelogram is a rhombus.

    c. Every square is a parallelogram.

    d. Every parallelogram is a square.

    e. Every rectangle is a square.

    f. Every square is a rectangle.

    g. Every rhombus is a square.

    h. Every square is a rhombus.

    i. A rhombus is equilateral and equiangular.

    j. A square is both equilateral and equiangular.

2. Lisa Steingart is an interior designer who writes a decorating column in a local newspaper. She advises people who are building houses to construct rectangular rather than square rooms. Why is it easier to decorate a rectangular room?

4-7.3 *Diagonals in Quadrilaterals*

STUDY GUIDE A *diagonal* is the line connecting the opposite vertices of a quadrilateral.

In quadrilateral *ABCD*:

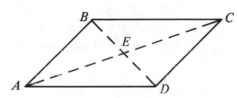

$\overline{AC}$ and $\overline{BD}$ are diagonals.

In parallelogram *ABCD*:

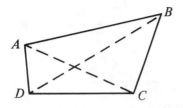

the diagonals $\overline{BD}$ and $\overline{AC}$ bisect each other at point *E*. $\overline{AE} = \overline{EC}$ and $\overline{BE} = \overline{ED}$.

In rhombus *EFGD*:

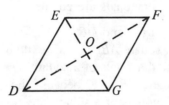

the diagonals bisect each other at *O* and are perpendicular to each other. $\overline{EO} = \overline{OG}$, $\overline{DO} = \overline{OF}$, and $\overline{EG} \perp \overline{DF}$. (The symbol for perpendicular is $\perp$.)

In rectangle *RSTE*:

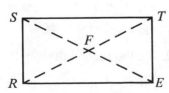

the diagonals bisect each other and are equal. $\overline{SF} = \overline{FE}$, $\overline{RF} = \overline{FT}$, and $\overline{SE} = \overline{RT}$.

In square *ESTO*:

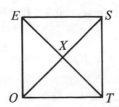

the diagonals bisect each other, are equal, and are perpendicular. $\overline{EX} = \overline{XT}$, $\overline{SX} = \overline{XO}$, $\overline{ET} = \overline{OS}$, and $\overline{ET} \perp \overline{OS}$.

1. The diagonals of a parallelogram _____ each other.

2. A diagonal of a quadrilateral connects the opposite _____ .

3. The diagonals of a square _____ each other and are

    _____ .

4. The diagonals of a _____ are perpendicular.

5. The diagonals of a _____ and a _____ are equal.

MODEL PROBLEM

Find the length of the line segments in each figure.

a.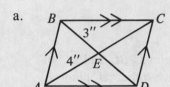

Solution: Find $\overline{EC}$, $\overline{ED}$, $\overline{BD}$, and $\overline{AC}$.
In parallelogram $ABCD$, $EC = 4''$
and $ED = 3''$ because the diagonals
of a parallelogram bisect each other.
$BD = 6''$ $(3'' + 3'')$.
$AC = 8''$ $(4'' + 4'')$.

b.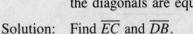

Solution: Find $\overline{EB}$ and $\overline{DC}$.
In rectangle $ACBD$, $EB = 5''$
because the diagonals of a rectangle
bisect each other. $DC = 10''$ because
the diagonals are equal.

c.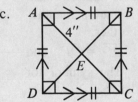

Solution: Find $\overline{EC}$ and $\overline{DB}$.
In square $ABCD$, $EC = 4''$ because
the diagonals of a square bisect each
other. $DB = 8''$ because the
diagonals of a square are equal.

EXERCISES

1. Draw and label the diagram for each problem below.

    a. In rectangle $ABCD$, if one diagonal equals $8''$, the second diagonal

        is __".

    b. In parallelogram $ABCD$, if $\overline{AC}$ and $\overline{BD}$ intersect at E and $AE = 4\frac{1}{2}''$,
        find EC.

    c. In rectangle $ABCD$, diagonals $\overline{AC}$ and $\overline{BD}$ intersect at E. If $AE = 9''$,
        find EC and BD.

    d. In rhombus $ABCD$, if the diagonals intersect at E, how many degrees
        are there in angle AEB?

2. 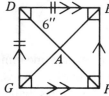 In square *DEFG*:

    a. Find $\overline{AF}$.

    b. Find $\overline{EG}$.

3. 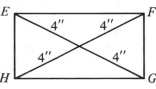 Is *EFGH* a parallelogram, rhombus, or rectangle?

4. Which properties are true for each figure named at the left?

	Diagonals Bisect Each Other	Diagonals Are Equal	Diagonals Are Perpendicular
a. Quadrilateral			
b. Parallelogram			
c. Rectangle			
d. Rhombus			
e. Square			

EXTRA FOR EXPERTS

1. When geometric figures are defined, they are arranged in order of their common properties. Remembering the definitions, rearrange these terms with the most general category first: rhombus, square, quadrilateral, parallelogram, rectangle, trapezoid.

2. Find the diagonal of a rectangle if the sides are 6″ and 8″.

3. List three words that have the prefix "quad-," and write their definitions.

4. If a diagonal of a square equals 6″, find the length of the side of the square.

5. A baseball diamond has 90 ft. as its distance from home plate to first base. Find the distance from home plate to second base.

4-8.1 *Polygons of More than Four Sides*

STUDY GUIDE The word "polygon" comes from Greek words meaning many angles. A *polygon* is a closed figure with lines as sides. Recall that a triangle is a polygon having three sides. A baseball diamond is an example of a quadrilateral, a four-sided polygon.

Other polygons are frequently seen: the outline of the Pentagon Building near Washington, D.C., is a five-sided polygon; the interiors of many socket wrenches are formed in the shape of six-sided polygons; and the familiar STOP sign has the shape of an eight-sided polygon.

Polygon	Name	Number of Sides
	Triangle	3
	Quadrilateral	4
	Pentagon	5
	Hexagon	6
	Octagon	8
Seven-sided	Heptagon	7
Nine-sided	Nonagon	9
Ten-sided	Decagon	10
Twelve-sided	Dodecagon	12

Equilateral polygons have all sides equal. Equiangular polygons have all angles equal. A *regular polygon* is both equilateral and equiangular.

Equilateral	**Equiangular**	**Regular**

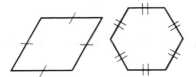

FACT FINDING

1. A polygon is a closed figure with _____ as sides.

2. A polygon that has eight sides is called an _____ .

3. A decagon has _____ sides.

4. A polygon with all sides equal is an _____ polygon.

5. A polygon with all angles equal is an _____ polygon.

6. A regular polygon has all _____ and all _____ equal.

7. A STOP sign has _____ sides and may be classified as a

   _____ .

MODEL PROBLEM

Determine whether the figure is a polygon, name it, and classify it as equilateral, equiangular, or regular.

a.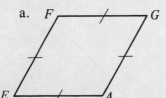

Solution: *EFGA* is a *polygon* because it is a closed figure with lines as sides.
EFGA is a *quadrilateral* because it is a polygon with four sides.
EFGA is *equilateral* because it has four equal sides.

b.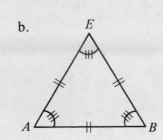

Solution: *AEB* is a *polygon*.
AEB is a *triangle*.
AEB is a *regular polygon* because all sides and all angles are equal.

1. Determine whether the figure is a polygon, name it, and classify it as regular, equilateral, or equiangular.

a.

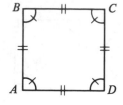

g.

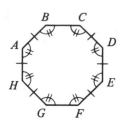

b.

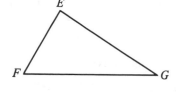

h.

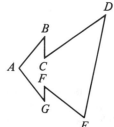

c.

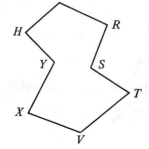

i.

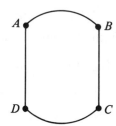

d.

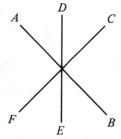

j.

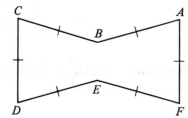

e.

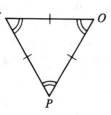

k.

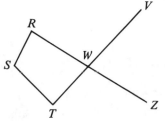

f.

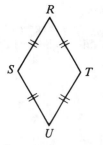

l.

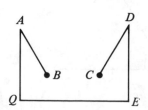

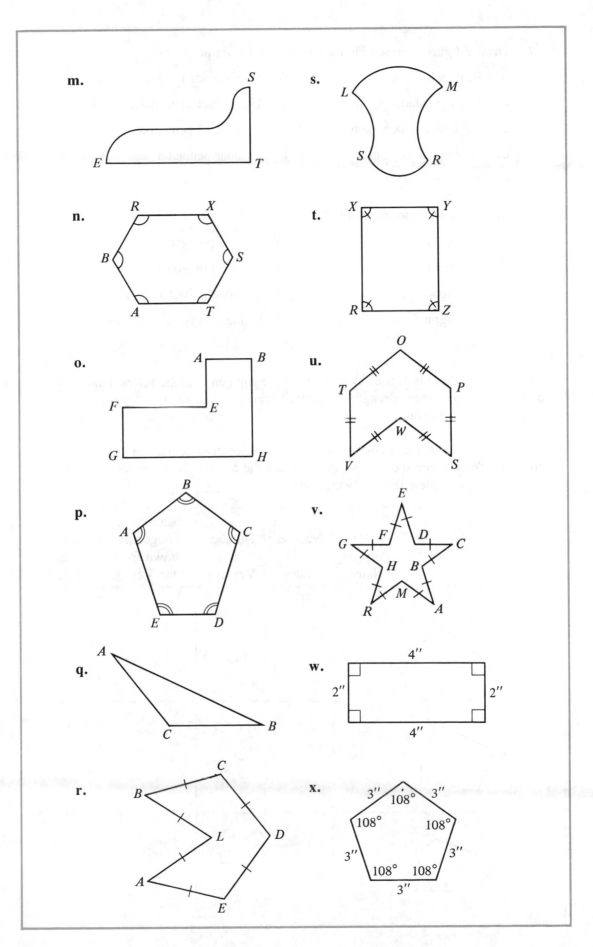

2. Draw a figure representing each of the following polygons:

a. Pentagon f. Equiangular hexagon

b. Regular quadrilateral g. Equilateral quadrilateral

c. Equiangular pentagon h. Equilateral pentagon

d. Equilateral heptagon i. Regular pentagon

e. Regular triangle

3. Match the letter to the number.

a. Dodecagon (1) 6-sided polygon

b. Octagon (2) 8-sided polygon

c. Decagon (3) 12-sided polygon

d. Hexagon (4) 9-sided polygon

e. Nonagon (5) 10-sided polygon

4. Arrange these polygons in order, with the polygon with the fewest number of sides first: pentagon, decagon, triangle, hexagon, dodecagon, octagon, quadrilateral, heptagon.

5. A *diagonal* is a line connecting two nonconsecutive vertices in a polygon. To compare the number of sides and the number of diagonals a polygon has, complete the following chart:

	Polygon Name	Number of Sides	Number of Vertices	Number of Diagonals Drawn from One Vertex	Total Number of Diagonals
a. 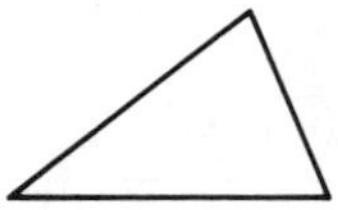					
b. 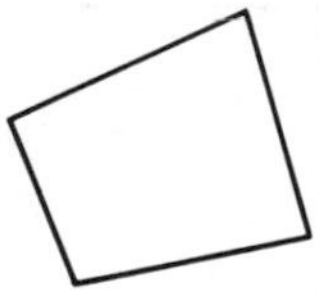					
c. 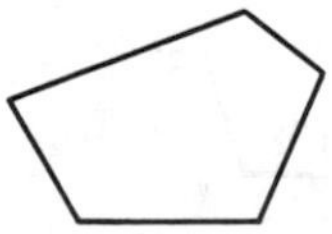					

d.
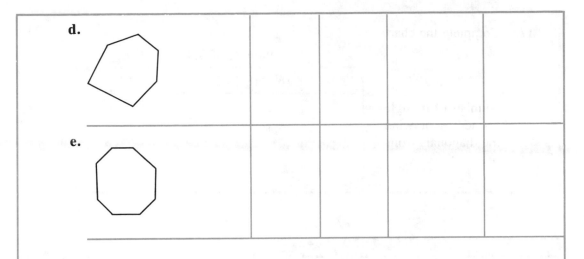

e.

6.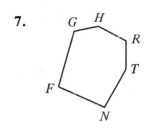

a. Name the diagonals of *RVTSP* that can be drawn:

   (1) From *R*   (2) From *P*

b. How many diagonals can be drawn from one vertex in this pentagon?

c. How many triangles are formed by drawing the diagonals from one and only one vertex?

7.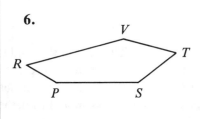

a. Name the diagonals of *GHRTNF* that can be drawn:

   (1) From *G*   (2) From *H*

b. How many triangles are formed by drawing the diagonals from *G*?

8. For each of the following polygons, find the *number* of triangles formed by drawing diagonals from one vertex:

   a. 3-sided polygon   **e.** 7-sided polygon

   b. 4-sided polygon   **f.** 8-sided polygon

   c. 5-sided polygon   **g.** 9-sided polygon

   d. 6-sided polygon   **h.** 10-sided polygon

9. A polygon of *n* sides can be divided into *n* − 2 triangles by drawing all the diagonals from one vertex. If a polygon has 15 sides, how many triangles can it be divided into by drawing the diagonals from one vertex?

10. How many triangles can a polygon *ABC* . . . having 20 sides be divided into by drawing the diagonals from *A*?

11. Complete the chart:

Number of Sides

13	14	16	18	22	24	30	17	x

Number of triangles formed by drawing the diagonals from one vertex

4-8.2 *Angles of Polygons*

STUDY GUIDE

The sum of the angles of a triangle is 180°. The sum of the angles of a polygon is found by counting the number of triangles formed by drawing all the diagonals from one vertex. If a polygon has eight sides, the sum of the angles is $(8 - 2)(180°)$ or $6(180°)$. From the results of Exercises 8 and 9 in the preceding section, if a polygon has n sides, the sum of the angles is $(n - 2)180°$.

A regular polygon is equiangular. To find the number of degrees in *one* angle of a regular polygon, divide the sum of the angles by the number of angles.

FACT FINDING

1. The sum of the angles of a triangle is _____°.

2. If a polygon has five sides, the sum of the angles is

    _____ × 180°.

3. To find the number of degrees in one angle of a regular polygon,

    _____ the sum of the angles by the number of angles.

MODEL PROBLEM

Name the regular polygon, and find the sum of the angles and the measure of each interior angle.

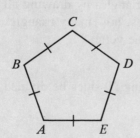

ABCDE is a regular pentagon.

The sum of the angles = $(5 - 2)\,180°$.

$$= 3(180)°.$$
$$= 540°.$$

Each angle = $\dfrac{\text{sum of the angles}}{\text{number of equal angles}}$.

$$= \dfrac{540°}{5} = 108°.$$

1. In a regular quadrilateral, find the sum of the angles and the size of one angle.

2. In a regular triangle, find the size of one angle.

3. In a decagon, find the sum of the angles.

4. In a regular decagon, find the size of one angle.

5. In a dodecagon, find the sum of the angles.

6. In a regular dodecagon, find the size of one angle.

7. The sum of the angles of a polygon is 1800°. The polygon has (a) 10 (b) 8 (c) 12 sides.

8. If the sum of the angles of a polygon is 3600°, how many sides does it have?

9. Complete the table for *regular* polygons.

	a.	b.	c.	d.	e.
Number of sides	8	5			
Sum of the angles			1260°		900°
Each angle				135°	

10. If four angles of a pentagon have a sum of 500°, what is the measure of the fifth angle? Is it a regular pentagon?

11. If three angles of a quadrilateral have a sum of 270°, what is the measure of the fourth angle? Is it a regular quadrilateral?

12. If nine angles of a decagon have a sum of 1300°, what is the measure of the tenth angle? Is it a regular decagon?

EXTRA FOR EXPERTS

1. Find the number of sides of a regular polygon each of whose interior angles contains:

    a. 90°   **b.** 120°   **c.** 150°

2. As the number of sides of a regular polygon increases, what change takes place in the number of degrees contained in each interior angle of the polygon?

3. The face of a honeycomb consists of interlocking regular hexagons.

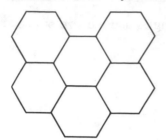

Can you interlock STOP signs as you can hexagons? Explain your answer, using the measures of the angles.

4. Can you tile a kitchen floor with regular pentagons? Explain your answer, using the measures of the angles.

4-9 *Points and Lines in a Circle*

STUDY GUIDE In the preceding sections, drawings of triangles and other polygons have been shown. If a dog is tied to a pole by a leash and walks around the pole, keeping the leash taut, what path does he follow as he walks? A circle is the correct answer.

A *circle* is the set of all points in a plane that are at the same distance, called the radius, from a fixed point, called the center, in the plane. The symbol for circle is ⊙.

To name a circle, we place a capital letter, such as *O*, near the center of the circle. The circle is then named "circle *O*" or "⊙*O*."

The *radius* of a circle is a line segment joining the center and any point on the circle. The plural of "radius" is "radii." A point is on a circle if its distance from the center is equal to a radius. In a circle, all radii are equal.

A *chord* of a circle is a line segment joining any two points on the circle.

A *diameter* of a circle is a chord that passes through the center of the circle. In a circle, all diameters are equal.

An *arc* is a part of a circle that consists of two points and the set of points on the circle between them.

Examples: In the figure below: *AB* is an arc of ⊙*O*. Arc *AB* is written as $\overset{\frown}{AB}$.

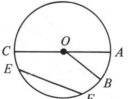

OB is a radius of ⊙*O*.
EF is a chord of ⊙*O*.
CA is a diameter of ⊙*O*.

A *sector* is part of a circle bounded by two radii and an arc.

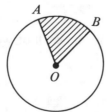

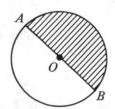

Sector *AOB* is part of the circle.

If *AB* is a diameter, sector *AOB* is half of circle *O*.

FACT FINDING

1. A _____ is a set of points at a fixed distance from a

_____ .

2. A line from the center of the circle to a point on the circle is called a

_____ .

3. In the same circle all radii are _____ .

4. A _____ connects two points on a circle.

5. The plural of "radius" is _____ .

6. We name a circle by using a capital letter at the _____ of the circle.

7. A diameter is a chord that passes through the _____ of the circle.

8. If *AB* and *CD* are diameters of the same circle, *AB* _____ *CD*.

MODEL PROBLEM

In ⊙*A*, name a radius, diameter, chord, and arc.

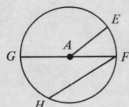

Solution: Radius = $\overline{AE}$.
Diameter = $\overline{GF}$.
Chord = $\overline{FH}$.
Arc = $\overparen{FH}$, $\overparen{GH}$, $\overparen{EG}$, $\overparen{EF}$.

EXERCISES

1. Mark a point *P* on a piece of paper. Take a ruler and find seven points, each 2 inches from point *P*. Call these points *A*, *B*, *C*, *D*, *E*, *F*, and *G*.

    a. What figure is suggested by the points *A*, *B*, *C*, *D*, *E*, *F*, *G*?

    b. Take your compass, and adjust it so that the sharp point is at *P* and the pencil point is at *A*. Draw a circle with center at *P*. Does the circle include the points *B*, *C*, *D*, *E*, *F*, and *G*?

    c. Name four radii of ⊙*P*.

    d. If *H* is a point 1 inch from point *P*, is *H* on the circle?

2. Construct ⊙*T* with a radius of $\frac{1''}{2}$. Name a radius.

3. Construct ⊙*R* with a diameter of 2″. Name a diameter.

4. In ⊙*E*: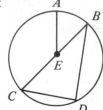

    a. Name three radii.

    b. Name a diameter.

    c. Name three arcs.

5. In ⊙*A*: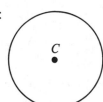

    a. Draw and name a radius.

    b. Draw and name a diameter.

    c. Draw and name a chord.

    d. Draw and name an arc.

6. In ⊙*B*: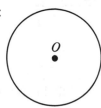

    a. Name three radii.

    b. Name a diameter.

    c. Name two chords.

    d. Name two arcs.

7. In ⊙*C*: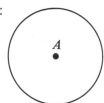

    a. Draw radius $\overline{CB}$.

    b. Draw diameter $\overline{AD}$.

    c. Draw chord $\overline{AB}$.

8. In ⊙*O*: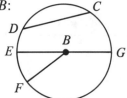

    a. Draw a radius $\overline{OX}$.

    b. How many radii can be drawn in circle *O*?

    c. What is true of the lengths of all the radii that can be drawn in ⊙*O*?

9. In ⊙*C*, if *AC* is a radius of 5″, how long must radius *CD* be?

10. In ⊙*D*: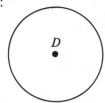

    a. Draw a diameter, and name it $\overline{EF}$.

    b. Draw another diameter, and name it $\overline{GH}$.

    c. How many diameters can be drawn in ⊙*D*?

    d. What is true of the lengths of all the diameters that can be drawn in ⊙*D*?

11. In ⊙*E*, if *GH* is a diameter of 9″, how long must diameter *LP* be?

12. Use a compass to draw a circle whose radius is .5 in. How long is a diameter in this circle?

13. Use a compass to draw a circle whose radius is 1.2″. How long is a diameter in this circle?

14. How does the length of a diameter of a circle compare to the length of the radius?

15. Complete the chart:

	a. ⊙A	**b.** ⊙B	**c.** ⊙C	**d.** ⊙D	**e.** ⊙E	**f.** ⊙F	**g.** ⊙G
Radius	3″	5.3″	$\frac{2''}{3}$	$1\frac{1}{2}'$			
Diameter					7″	16′	4.8″

16. Find the diameter of a circle whose radius is:

a. 4 cm

b. $\frac{4}{7}$ inch

c. 6.4 feet

17. If a circle has a radius of r units, what is the length of the diameter of the circle in terms of r?

18. If a circle has a diameter of d units, what is the length of the radius of the circle in terms of d?

19. TRUE or FALSE?

    a. A circle is a closed figure.

    b. A diameter is a chord.

    c. A diameter of a circle is one-half as long as a radius of the same circle.

    d. A circle has exactly two radii.

20. Construct circle D with the radius = _____,
$$p''$$

    with the diameter = _____.
$$P''$$

21. Construct circle E with a radius of 2 cm.

22. 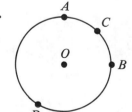 In ⊙O, $\overarc{AC} = \overarc{CB}$.

    a. If $\overarc{AC} = 3''$, find the length of $\overarc{CB}$.

    b. If $\overarc{AD} = 8''$ and $\overarc{DB} = 6''$, find the lengths of $\overarc{ADB}$ and $\overarc{CBD}$.

23.

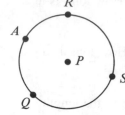

In $\odot P$, if we name an arc $\overset{\frown}{AS}$, it is not clear whether we mean arc $\overset{\frown}{ARS}$ or $\overset{\frown}{AQS}$. Therefore it is better to use three letters to name an arc and thus avoid confusion.

a. If $\overset{\frown}{AR} = 2.6''$ and $\overset{\frown}{RS} = 3.1''$, find the length of $\overset{\frown}{ARS}$.

b. If $\overset{\frown}{AQ} = 4\dfrac{1}{2}''$ and $\overset{\frown}{QS} = 5\dfrac{1}{2}''$, find the length of $\overset{\frown}{AQS}$.

24.

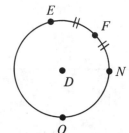

$\overset{\frown}{EF} = \overset{\frown}{FN}$, $\overset{\frown}{OE} = 9.8$ cm, $\overset{\frown}{EF} = 2.3$ cm.

a. Find the length of $\overset{\frown}{OEF}$.

b. If $\overset{\frown}{NO} = 4.5$ cm, find the length of $\overset{\frown}{FNO}$.

25.

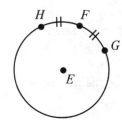

$\overset{\frown}{FH} = 2\dfrac{1}{2}''$, $\overset{\frown}{FG} = \overset{\frown}{FH}$.
Find the length of $\overset{\frown}{FG}$.

26.

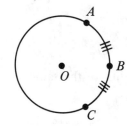

$\overset{\frown}{AB} = \overset{\frown}{BC}$, $\overset{\frown}{AB} = \dfrac{3''}{5}$.
Find the length of $\overset{\frown}{ABC}$.

27.

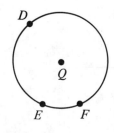

$\overset{\frown}{DF} = 9.8''$, $\overset{\frown}{EF} = 2.9''$.
Find the length of $\overset{\frown}{DE}$.

28.

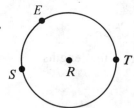

$\overset{\frown}{ES} = 3.6''$, $\overset{\frown}{ST} = 2(\overset{\frown}{ES})$.
Find the lengths of $\overset{\frown}{ST}$ and $\overset{\frown}{EST}$.

29.

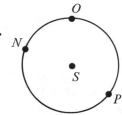

$\widehat{NO} = x''$, $\widehat{OP} = 2(\widehat{NO})$.
Find the length of $\widehat{OP}$ in
terms of x.

30. For each circle O, name the shaded sector.

a. 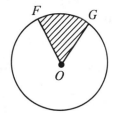 b. 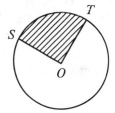 c.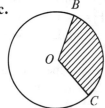

31. Indicate whether the shaded area for each circle O is a sector. If not, tell why it is not.

a. 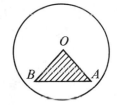 d.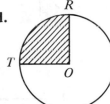

b. 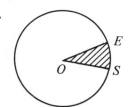 e.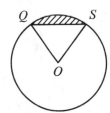

c.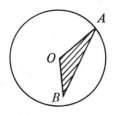

32. For Arthur Jones, a student, the day is divided into the following activities:

a. Which activity forms the largest sector?

b. Which activity forms the smallest sector?

c. Is Arthur typical of students you know?

1. Draw a circle. Without changing your compass setting, mark off an arc *AB*. Beginning at *B*, mark off a second arc equal to the first and call it *BC*. Continue until you get back to point *A*. Join the crossing points on the circle. What figure did you draw?

2. The figure in Extra 1 is an inscribed polygon. A polygon inscribed in a circle has its vertices on the circle. From the figure for Extra 1, inscribe a regular triangle in the same circle.

3. An inscribed polygon has its vertices on the circle. Which of the diagrams below illustrate inscribed polygons?

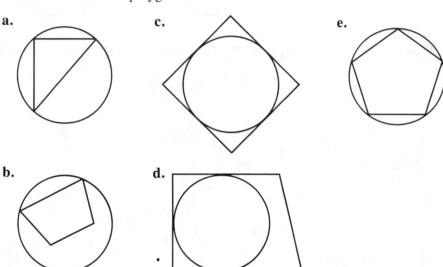

a.

c.

e.

b.

d.

.

4-10 *Perimeter of a Polygon*

STUDY GUIDE   The distance around a polygon is the *perimeter* of the polygon. To calculate the perimeter, the lengths of the sides of the polygon are added together.

In a *triangle* the perimeter is found by adding the lengths of the three sides:

$$P = a + b + c.$$

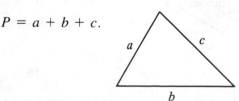

In a *rectangle* the opposite sides are equal, so the perimeter is:

$$P = a + a + b + b \quad \text{or} \quad 2a + 2b.$$

In a *square*, since all four sides are equal, the perimeter is:

$$P = s + s + s + s \quad \text{or} \quad P = 4s.$$

FACT FINDING

1. The distance around a polygon is called the _____.

2. The perimeter is equal to the _____ of the lengths of the sides of a polygon.

3. A square has four _____ sides.

4. A _____ has three sides.

5. The opposite sides of a rectangle are _____ to each other.

6. To calculate the perimeter of a polygon we find the _____ of the lengths of the sides.

MODEL PROBLEMS

1. Find the perimeter of the triangle.

   Solution: $P = a + b + c$
   $= 4 + 3 + 6$
   $= 13.$

2. Find the perimeter of the square whose side is 5.

   Solution: $P = 4s$
   $= 4 \times 5$
   $= 20.$

3. Find the perimeter of the rectangle whose length is 6 and whose width is 3.

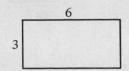

   Solution: $P = 2L + 2W$
   $= 2 \times 6 + 2 \times 3$
   $= 12 + 6$
   $= 18.$

1. Find the perimeters of the figures below.

a.

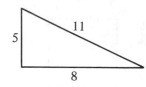

f.

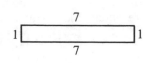

k.

b.

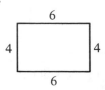

g.

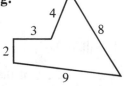

l.

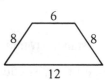

c.

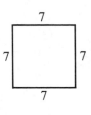

h.

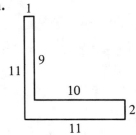

m.

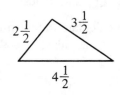

d.

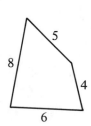

i.

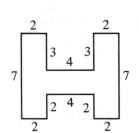

n.

e.

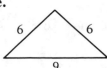

j.

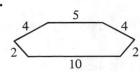

o.

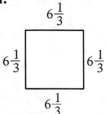

2. Find the perimeters of the triangles the lengths of whose sides are given below.

	a.	b.	c.	d.	e.	f.	g.	h.	i.	j.
r	3	$2\frac{1}{2}$	3.2	3.1	2.9	$6\frac{1}{3}$	$5\frac{1}{4}$	$5\frac{3}{4}$	23.8	$3\frac{1}{4}$
t	4	$4\frac{1}{2}$	4.7	3.1	2.9	$4\frac{1}{2}$	$2\frac{1}{3}$	$11\frac{7}{8}$	15.5	$5\frac{1}{2}$
m	6	$5\frac{1}{2}$	3.2	3.1	4.3	$7\frac{1}{2}$	$4\frac{1}{2}$	$15\frac{1}{3}$	18.9	$2\frac{1}{2}$
P										

3. Find the perimeters of the polygons the lengths of whose sides are given below.

	a.	b.	c.	d.	e.	f.	g.	h.	i.	j.
w	10	15	4.9	$4\frac{7}{8}$	6.2	12.6	$8\frac{3}{8}$	$2\frac{1}{2}$	$4\frac{4}{5}$	$6\frac{1}{2}$
x	4	7	6.3	$5\frac{5}{8}$	6.2	5.3	$4\frac{1}{8}$	$3\frac{1}{2}$	$4\frac{1}{3}$	$4\frac{3}{4}$
y	10	15	4.9	$2\frac{9}{16}$	6.2	13.7	$5\frac{5}{16}$	$7\frac{1}{3}$	$7\frac{1}{2}$	$8\frac{1}{2}$
z	4	7	6.3	$7\frac{3}{4}$	6.2	15.9	$4\frac{1}{4}$	$9\frac{1}{3}$	$9\frac{2}{3}$	$10\frac{1}{3}$
P										

4. Find the perimeter for each square the length of whose side is given in the table below. (*Hint:* $P = 4s$.)

	a.	b.	c.	d.	e.	f.	g.	h.	i.	j.
s	5	11	4.3	3.2	6.8	$4\frac{1}{2}$	$8\frac{2}{3}$	$6\frac{3}{8}$	$4\frac{7}{8}$	$5\frac{3}{4}$
P										

5. Find the perimeters of the rectangles the lengths of whose two sides are given below.

	a.	b.	c.	d.	e.	f.	g.	h.	i.	j.
L	3.2	7.1	6.9	$4\frac{1}{2}$	$6\frac{2}{3}$	$5\frac{5}{8}$	$3\frac{4}{5}$	$7\frac{7}{12}$	$9\frac{3}{5}$	$4\frac{1}{3}$
W	4.6	11.7	10.7	$6\frac{1}{3}$	$4\frac{3}{4}$	$9\frac{7}{16}$	$4\frac{1}{2}$	$9\frac{3}{4}$	$10\frac{3}{4}$	$9\frac{5}{16}$
P										

6. Solve the following problems:

   a. How much fencing is needed to enclose a vegetable garden 7 feet long and $6\frac{1}{2}$ feet wide?

   b. How much molding is needed to go around the base of a room $14\frac{1}{2}$ feet wide by $17\frac{1}{3}$ feet long, if a doorway 3 feet wide is excluded?

EXTRA FOR EXPERTS

1.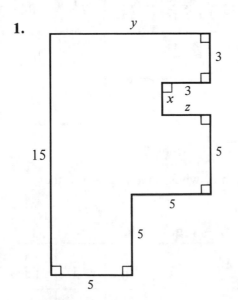

Find the values of x, y, and z, and then calculate the perimeter.

2. A fence is needed around a swimming pool. The pool is 15 feet long by 12 feet wide and is surrounded by an 8-foot-wide cement walk. What is the length of the fencing needed to enclose the pool and walk?

3. What is the length of each side of a square if 40 yards of cyclone fencing is needed to enclose it?

4-11 *Perimeter of a Polygon (Continued)*

STUDY GUIDE As you have learned, triangles are classified by the lengths of their sides and by the measures of their angles. Triangles are equilateral, isosceles, or scalene. An equilateral triangle has three equal sides, while an isosceles triangle has two equal sides and a scalene triangle has no equal sides.

If the perimeter of a polygon and the lengths of all sides except one are known, we can find the length of the remaining side. We add the known sides, and subtract this sum from the perimeter.

FACT FINDING

1. An equilateral triangle has _____ equal sides.

2. An _____ triangle has two equal sides.

3. Triangles are classified by the _____ of their sides and by the _____ of their angles.

4. To find the missing length of a side of a triangle, the _____ of the sides is _____ from the perimeter.

MODEL PROBLEM

Find the length of the third side of an isosceles triangle if the length of one of the equal sides is $6\frac{1}{3}$ and the perimeter is $20\frac{2}{3}$.

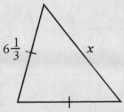

Solution: If one of the equal sides is $6\frac{1}{3}$,

then the other equal side is also $6\frac{1}{3}$.

Thus $6\frac{1}{3} + 6\frac{1}{3} = 12\frac{2}{3}$. The perimeter equals the sum of the sides, so:

$$20\frac{2}{3} = 6\frac{1}{3} + 6\frac{1}{3} + x.$$

$$20\frac{2}{3} = 12\frac{2}{3} + x$$

$$8 = x.$$

1. Find the missing side in each of the following isosceles triangles:

a.

$$P = 11$$

$$x = \_\_$$

d.

$$P = 7$$

$$x = \_\_$$

g.

$$P = 16\frac{3}{4}$$

$$c = \_\_$$

b.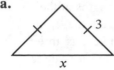

$$P = 8.7$$

$$y = \_\_$$

e.

$$P = 25$$

$$y = \_\_$$

h.

$$P = 24\frac{3}{4}$$

$$d = \_\_$$

c.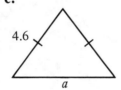

$$P = 14$$

$$a = \_\_$$

f.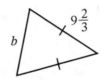

$$P = 27$$

$$b = \_\_$$

i.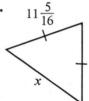

$$P = 32\frac{7}{8}$$

$$x = \_\_$$

2. Complete the following table. The length given is one of the equal sides of an isosceles triangle.

	a.	b.	c.	d.	e.	f.	g.	h.	i.	j.
Equal side	$7\frac{1}{2}$	$4\frac{3}{5}$	$11\frac{2}{3}$	$6\frac{1}{4}$	$5\frac{4}{5}$	3.7	4.9	8.2	16.7	40.3
P	$22\frac{2}{3}$	$15\frac{2}{3}$	$32\frac{5}{6}$	$18\frac{1}{2}$	$14\frac{3}{16}$	12.9	13	27.1	39.8	110.7
3rd side, x										

3. For the following equilateral triangles, one side is given. Find the perimeter.

a. $s = 4$

f. $s = 11.7$

k. $s = 19\frac{1}{3}$

p. $s = 8\frac{3}{4}$

b. $s = 9$

g. $s = 2.3$

l. $s = 3\frac{1}{2}$

q. $s = 11\frac{1}{2}$

c. $s = 3.1$

h. $s = 50.6$

m. $s = 4\frac{2}{7}$

r. $s = 14\frac{2}{3}$

d. $s = 4.7$

i. $s = 70.9$

n. $s = 16\frac{3}{16}$

s. $s = 4y$

e. $s = 5.2$

j. $s = 111.16$

o. $s = 9\frac{4}{5}$

t. $s = 2x$

4. For the following equilateral triangles, the perimeter is given. Find the length of a side of the triangle.

a. $P = 36$

d. $P = 64$

g. $P = 435$

i. $P = 57.9$

b. $P = 54$

e. $P = 18$

h. $P = 120.9$

j. $P = 41.1$

c. $P - 57$

f. $P = 34\frac{1}{5}$

5. Find the length of the side of each square whose perimeter is given.

a. $P = 20$

d. $P = 49.45$

g. $P = 99.3$

i. $P = 112$

b. $P = 34$

e. $P = 85.6$

h. $P = 44.9$

j. $P = 3\frac{3}{4}$

c. $P = 117$

f. $P = 77$

6. Find the missing side in each of the triangles below.

a.

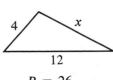

$P = 26$

$x = $ ___

b.

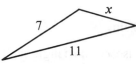

$P = 27$

$x = $ ___

c.

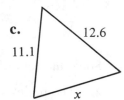

$P = 34.2$

$x = $ ___

d.

$P = 13$

$x = $ ___

e.

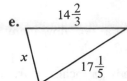

$P = 49\frac{3}{5}$

$x = $ ___

f.

$P = 37\frac{3}{4}$

$x = $ ___

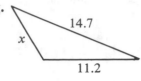

g. 14.7 x 11.2

h. 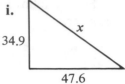 10.5 x 17.3

i. 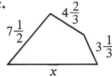 34.9 x 47.6

$P = 29.6$

$P = 36.1$

$P = 111.9$

$x = \underline{\quad}$

$x = \underline{\quad}$

$x = \underline{\quad}$

7. In each of the polygons below, find the missing side.

a.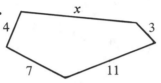
4 x 3 7 11

b.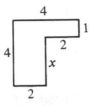
4 1 2 x 2

c.
$7\frac{1}{2}$ $4\frac{2}{3}$ $3\frac{1}{3}$ x

$P = \underline{\quad}$

$P = 16$

$P = 26\frac{2}{5}$

$x = 16$

$x = \underline{\quad}$

$x = \underline{\quad}$

d.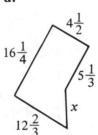
$4\frac{1}{2}$ $16\frac{1}{4}$ $5\frac{1}{3}$ x $12\frac{2}{3}$

e.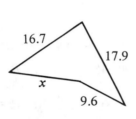
16.7 17.9 x 9.6

f.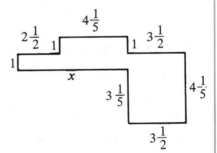
$2\frac{1}{2}$ 1 $4\frac{1}{5}$ 1 $3\frac{1}{2}$ x $3\frac{1}{5}$ $4\frac{1}{5}$ $3\frac{1}{2}$

$P = 50$

$P = 54.3$

$P = 34\frac{2}{3}$

$x = \underline{\quad}$

$x = \underline{\quad}$

$x = \underline{\quad}$

8. Solve the following problems:

   a. The perimeter of an isosceles triangle is 40 feet. What is the length of the two equal sides if the third side is 10 feet?

   b. The perimeter of a square is 24 feet. What is the length of each side?

   c. At $0.69 a yard for ribbon, how much will it cost to bind a rug 6.5 yards by 9.2 yards?

   d. If fencing costs $2.10 a foot, how many feet of fencing can be purchased for $88.00?

   e. A square picture needs 33 inches of framing. How long is each side?

1. If each side of a 9-foot equilateral triangle is tripled, what is the length of each new side? What is the perimeter?

2. How many begonia plants are needed for the border of a triangular garden that is 7 feet by 9 feet by 8 feet if the begonias are planted 1 foot apart?

4-12 *Circumference of a Circle*

STUDY GUIDE The *circumference* of a circle is the distance around the circle.
The *radius* of a circle is a line segment joining the center of the circle and any point on the circle.

The *diameter* of a circle is a chord that passes through the center of the circle and has both endpoints on the circle. The radius of a circle is one-half the diameter of the circle.

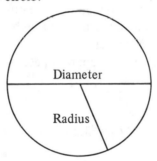

FACT FINDING

1. The distance around the circle is the _____.

2. The radius of a circle is one-half the _____ of the circle.

3. The _____ is a line segment that has its endpoints on the circle and passes through the center of the circle.

4. The _____ of a circle has one endpoint on the circle and one endpoint at the center of the circle.

EXERCISES

1. Using a piece of string and a ruler, measure the circumferences of the following objects:

    a. A soda can   **b.** A record album   **c.** A one-pound coffee can

2. Complete the following chart by dividing each diameter into the circumference:

Circumference	Diameter	$\dfrac{\text{Circumference}}{\text{Diameter}}$
22′	7′	
15.7′	5′	
34.54′	11′	

4-13 *Circumference of a Circle (Continued)*

STUDY GUIDE

The *circumference* of a circle is the distance around a circle. In Exercise 2 of the preceding section, the answer found when each diameter is divided into the circumference is called *pi* (π). The circumference is calculated by multiplying π times the diameter (d). Pi is approximately equal to 3.14.

When π is written as an approximate fraction, it is expressed as $3\frac{1}{7}$ or $\frac{22}{7}$. Therefore

$$C = \pi d \quad \text{or} \quad C = 3.14 \times d \quad \text{or} \quad C = \frac{22}{7} \times d.$$

FACT FINDING

1. The distance around a circle is called the _____ .

2. The diameter of a circle is represented by _____ .

3. Circumference is calculated by _____ _____ times d.

4. Pi is approximately equal to 3.14 or _____ .

MODEL PROBLEM

Find the circumference of the circle when $d = 6$ cm.

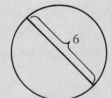

Solution: $C = 3.14 \times d$
$ = 3.14 \times 6$
$ = 18.84$ cm.

1. Find the circumference of each circle whose diameter is given below.

    a. 7"   f. 13"   k. 3.8 mm   p. 14"

    b. 3"   g. 3.14"   l. 17"   q. 4.3"

    c. 4"   h. 9.3"   m. 10.4 yd.   r. 3.3 m

    d. 10'   i. 5.2'   n. 17.6 m   s. 1.5"

    e. 2.1"   j. 48'   o. 9 ft.   t. 22'

2. Find the circumference of each circle whose radius is given below.
    (*Hint:* $d = 2r$; therefore $C = 2 \times 3.14 \times r$.)

    a. 63 mi.   f. 19.1 cm   k. 8.3"   p. 9.4 m

    b. 28'   g. 8.6'   l. 3.7 m   q. 44'

    c. 8.2"   h. 78.1"   m. 3.5 mi.   r. 96 yd.

    d. 7.7 m   i. 11.1 ft.   n. 17.6 m   s. 6.2"

    e. 2'   j. 1.5"   o. 3.1"   t. 11.7 ft.

3. Solve the following problems:

    a. What distance do you ride on one turn of a Ferris wheel when the
        seat is 20 feet from the center?

    b. How much fencing is needed for a circular garden with a diameter of
        11 feet? If the fencing is $3 a foot, what will the total cost be?

    c. The diameter of the earth is 7918 miles. What is the circumference?

d. A bicycle has a wheel with a diameter of 21 inches. How far will the wheel go in one complete turn? How many turns will the wheel need to take to go 3360 feet?

e. How many times must you run around a circular track with a diameter of 105 feet in order to run a mile?

f. A circular pond has a radius of 4.3 feet. What is the circumference?

EXTRA FOR EXPERTS

1. A pool has a diameter of 10 feet. To build a fence 8 feet from the edge all around the pool, how much fencing do you need?

2. If a circular track is 440 m all around, what is the radius of the track?

3. A flower bed that is a half circle with a radius of 3′ needs a fence.

3′

a. How much fencing will it require?

b. At $1.25 a foot, what will the fencing cost?

MODEL PROBLEMS

1. Find the circumference of a circle whose diameter is $4\frac{1}{2}$.

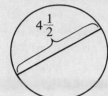

Solution: $C = \pi d$

$$= \frac{22}{7} \times 4\frac{1}{2}$$

$$= \frac{99}{7} \quad \text{or} \quad 14\frac{1}{7}.$$

2. Find the circumference of a circle whose $r = 3\frac{1}{4}$.

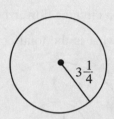

Solution: Diameter $= 2 \times$ radius

$$= 2 \times 3\frac{1}{4}$$

$$= 6\frac{2}{4} \quad \text{or} \quad 6\frac{1}{2}$$

$$d = \frac{13}{2}.$$

$$C = \pi d$$

$$= \frac{22}{7} \times \frac{13}{2} = \frac{286}{14}$$

$$= 20\frac{6}{14} \quad \text{or} \quad 20\frac{3}{7}.$$

EXERCISES

1. Calculate the circumferences of the circles whose diameters are given below.

a. $d = 3\frac{1}{2}$ m

b. $d = 5\frac{1}{2}$ ft.

c. $d = 8\frac{3}{4}$ m

d. $d = 4\frac{3}{4}$ m

e. $d = \frac{7}{8}$ mi.

f. $d = \frac{1}{2}$ ft.

g. $d = 6\frac{2}{3}''$

h. $d = 4\frac{2}{5}$ ft.

i. $d = 8\frac{3}{8}$ m

j. $d = 53\frac{1}{2}$ ft.

2. Find the circumferences of the following circles whose radii are given:

a. $r = 3\frac{1}{2}$ ft.

b. $r = 4\frac{1}{7}''$

c. $r = 8\frac{2}{3}$ ft.

d. $r = 21\frac{3}{4}''$

e. $r = 17\frac{1}{2}$ m

f. $r = 9\frac{1}{5}$ ft.

g. $r = 14\frac{1}{4}'$

h. $r = 50\frac{1}{2}''$

i. $r = 25\frac{1}{3}$ ft.

j. $r = 32\frac{3}{4}$ cm

k. $r = \frac{7}{8}$ m

l. $r = \frac{3}{4}$ m

3. Solve the following problems:

    a. (1) How far will a wheel go in one turn if $d = 7$ yards?

          (2) How far will it go in 3 turns?

          (3) How far will it go in 10 turns?

          (4) How many turns must it revolve in order to go 55 yards?

    b. Find the circumference of a wheel with a diameter of $21\frac{1}{2}''$.

    c. What is the circumference of a tire whose radius is $11\frac{1}{2}''$?

    d. Find how much fencing is needed to surround a circular flower bed with a diameter of $17\frac{1}{2}$ m. At \$0.99 a meter, what is the total cost of the fencing?

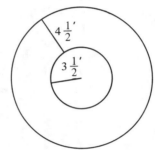

    e. Find the circumference of the larger circle if the radius of the inner circle is $3\frac{1}{2}'$ and the distance from the edge of the inner circle to the edge of the outer circle is $4\frac{1}{2}'$.

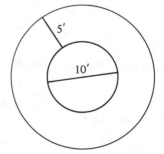

    f. Find the circumference of the outer circle.

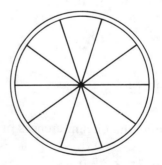

    g. Find the inside diameter of a bicycle wheel whose spoke is $10\frac{1}{2}''$.

    h. A semicircular flower bed needs a fence. The semicircle has a diameter of $10\frac{1}{2}$ feet. How much fencing is needed? At \$3.10 a foot, what is the cost of the fencing?

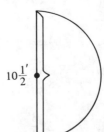

1.  A circular flower bed is planted around a
fountain. The fountain has a diameter of 6′.
The distance from the fountain to the outer
edge of the circle is 5′.

    a. What is the diameter of the larger circle?

    b. What is the circumference of the larger circle?

    c. What is the circumference of the fountain?

    d. What is the total length of fencing needed to go around the fountain
and around the flower bed?

    e. At $2.90 a foot, how much will the fencing cost?

2. Find the diameter of a barrel whose circumference is 44 cm.

4-14 *Finding Perimeters and Circumferences*

STUDY GUIDE The *perimeter* of a polygon is the distance around the polygon.
To calculate the perimeter, the lengths of the sides of the polygon
are added together.

The formula for the perimeter of a triangle is

$$P = s_1 + s_2 + s_3,$$

where s_1 = one side, s_2 = second side, and s_3 = third side.
If the triangle is equilateral, the formula is

$$P = 3s.$$

For the perimeter of a square the formula is

$$P = 4s.$$

The formula for the perimeter of a rectangle is

$$P = 2L + 2W, \quad \text{where } L \text{ is length and } W \text{ is width.}$$

The *circumference* of a circle is defined as the distance around the circle.
To calculate the circumference, the diameter is multiplied by π:

$$C = \pi d.$$

If the radius is given,

$$C = 2\pi r$$

because the diameter is twice the radius.

FACT FINDING

1. In a circle the radius is one-half the _____ .

2. If the diameter is known, the formula for the circumference of a circle is _____ .

3. Perimeter is defined as the distance _____ a polygon.

4. A _____ is a polygon with four equal sides and four right angles.

5. A rectangle has _____ sides equal to each other.

6. The formula for the perimeter of a _____ is given by $P = 2L + 2W$.

7. A triangle is a polygon with _____ sides.

8. The perimeter of a(n) _____ triangle is $P = 3s$.

9. The formula for the perimeter of a square is _____ .

MODEL PROBLEMS

1. Find the perimeter of a square whose side is $9\frac{1}{2}''$.

$9\frac{1}{2}''$

Solution: $P = 4s$

$= 4 \times 9\frac{1}{2}''$

$= 38''$.

2. Find the perimeter of a triangle whose sides are 15″, 11″, and 9″.

9″   11″

15″

Solution: $P = s_1 + s_2 + s_3$

$= 15'' + 11'' + 9''$

$= 35''$.

3. Find the perimeter of a rectangle whose $L = 9.2$ m and $W = 6.4$ m.

6.4 m

9.2 m

Solution: $P = 2L + 2W$

$= (2 \times 9.2 \text{ m}) + (2 \times 6.4 \text{ m})$

$= 18.4 \text{ m} + 12.8 \text{ m}$

$= 31.2 \text{ m}$.

4. Find the circumference of a circle whose $d = 11\frac{1}{2}''$.

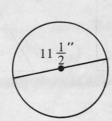

Solution: $C = \pi d$

$= \frac{22}{7} \times 11\frac{1}{2}$

$= \frac{22}{7} \times \frac{23}{2}$

$= \frac{253}{7} = 36\frac{1}{7}''$

$\left(\text{Use } \pi = \frac{22}{7} \text{ because the}\right.$
diameter is written in fraction
form.$\Big)$

EXERCISES

1. Find the perimeter or circumference of each of the following figures:

a.
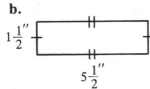
3.2″

f.
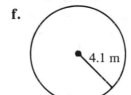
4.1 m

k.
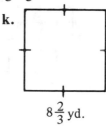
$8\frac{2}{3}$ yd.

b.
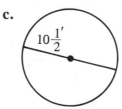
$1\frac{1}{2}''$
$5\frac{1}{2}''$

g.
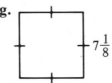
$7\frac{1}{8}$

l.
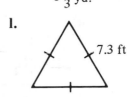
7.3 ft

c.
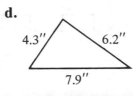
$10\frac{1}{2}'$

h.
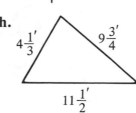
$4\frac{1}{3}'$ $9\frac{3}{4}'$
$11\frac{1}{2}'$

m.
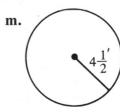
$4\frac{1}{2}'$

d.
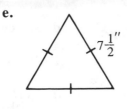
4.3″ 6.2″
7.9″

i.
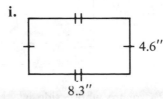
4.6″
8.3″

n.
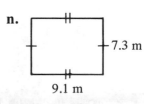
7.3 m
9.1 m

e.
7$\frac{1}{2}''$

j.
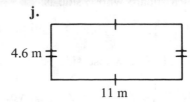
4.6 m
11 m

o.
$7\frac{2}{3}''$

2.

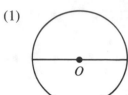 (1)   (2) 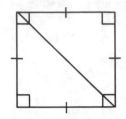   (3)

     a.   Which picture above has a diagonal?

     b.   Which picture above has a diameter?

     c.   Which picture above has a hypotenuse?

     d.   For which picture above does $P = 4s$?

     e.   For which picture above does $P = s_1 + s_2 + s_3$?

3.   Find the perimeters of the following polygons:

   a.   **d.**   **g.**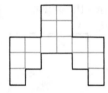

   b.   **e.**   **h.**

   c.   **f.** 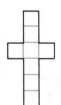   **i.**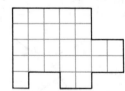

4.   Find the perimeter of a square whose side is:

     a.   54″   **b.**   15″   **c.**   30 m   **d.**   25 cm

5.   Find the perimeter of a square whose side is:

     a. $5\frac{1}{2}''$ **b.** $2\frac{1}{2}''$ **c.** $7\frac{3}{4}$ yd.   **d.** $59\frac{1}{4}''$

6. Find the perimeter of an equilateral triangle whose side is:

    a. 5″
    b. 7.2′
    c. $11\frac{1}{2}''$
    d. 6.9 cm

7. Find the perimeter of a square whose side is:

    a. 11.2″
    b. 14.6 m
    c. 19.6′
    d. 44.3 cm

8. Find the perimeter of a rectangle whose length and width are:

    a. $L = 9'$
    b. $L = 11''$
    c. $L = 9.3$ cm
    d. $L = 3\frac{1}{2}'$

       $W = 6'$
       $W = 14''$
       $W = 4.7$ cm
       $W = 4'$

9. Find the perimeter of a triangle whose sides are:

    a. 6′, 7′, 11′
       c. 8.4″, 8″, 9.2″

    b. $4\frac{1}{2}''$, $9\frac{1}{3}''$, $6\frac{2}{3}''$
    d. 15.4″, 19.8″, 19.6″

10. Find the circumference of a circle whose diameter is:

    a. 3′
    b. 4′
    c. 8.2″
    d. 19.6′

11. Find the circumference of a circle whose radius is:

    a. 11″
    b. 2.3 ft
    c. 8.6 cm
    d. 54.3 cm

12. Find the circumference of a circle whose diameter is:

    a. $11\frac{2}{3}'$
    b. $6\frac{3}{4}''$
    c. $14\frac{1}{4}'$
    d. $7\frac{1}{2}''$

13. Find the circumference of a circle whose radius is:

    a. $5\frac{2}{3}''$
    b. $9\frac{1}{2}'$
    c. $11\frac{3}{4}''$
    d. $16\frac{3}{8}$ cm

14. A circular flower bed has a diameter of 7.2 ft. How many feet of fencing are required to enclose the garden?

15. A rectangular room has a length of 14.1′, a width of 8.3′, and a single door 2.5′ wide. How much molding is needed to go around the baseboard in the room?

16. The perimeter of a square is 24.8′. What is the side of the square?

17. A rectangular pool 21′ by 18′ needs a walk around it that is 3′ wide. What are the length and the width of the rectangle, including the walk? What is the outside perimeter of the walk?

18. A circular flower bed has a radius of $11\frac{1}{2}$ feet. How much fencing is needed to enclose the flower bed?

19. What is the perimeter of a triangular plot of land whose sides are 18 feet, 21 feet, and 24 feet?

20. The distance through the center of a circular track is 20 yd. What is the distance a runner runs around the track?

21. A rectangular room is $19\frac{1}{2}$ feet long and $14\frac{3}{4}$ feet wide. Find the perimeter of the room. If the room has a door 3 feet wide, how much molding is needed to go around the baseboard? If the molding is $3.80 a foot, how much will it cost?

22. A square picture has a side of 19.2″. How much framing is needed for the picture?

23. An equilateral triangle has a side of 17.4″. What is the perimeter of the triangle?

EXTRA
FOR
EXPERTS

1. Find the perimeter of each polygon.

a.

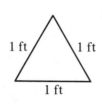 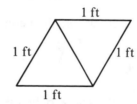 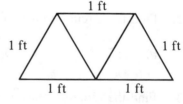

1 triangle   2 triangles   3 triangles

P = __   P = __   P = __

4 triangles   5 triangles

P = __   P = __

b.

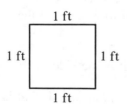 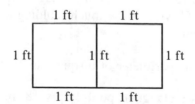

1 square   2 squares

P = __   P = __

3 squares   4 squares   5 squares   10 squares   25 squares

P = __   P = __   P = __   P = __   P = __

2. A man has a rectangular plot of land whose length is 14′ and whose perimeter is 54′. Find the length of the rectangle.

Measuring Your Progress

1. **a.** Identify each angle as an acute, obtuse, right, or straight angle.

    (1) 37°   (3) 180°   (5) 152°   (6) 69°

    (2) 110°   (4) 90°

  b. (1) Draw an acute angle, an obtuse angle, a right angle, and a straight angle.

      (2) Label each angle, and name each in at least three different ways.

      (3) Measure each angle with your protractor, and determine how closely you estimated the approximate size of your angle.

  c. Complete the following:

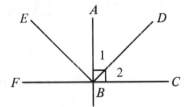

    (1) $\angle FBA = \angle ABE + \angle \underline{\ \ }$

    (2) $\angle ABC = 90° - \underline{\ \ }$

2. Classify each of the following triangles as scalene, isosceles, or equilateral, and as right, acute, or obtuse. Use two adjectives to describe each triangle.

  a.

  c.

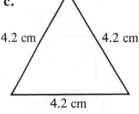

  b.

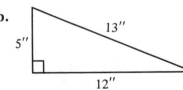

3. **a.** If $\angle A + \angle B$ in $\triangle ABC = 105°$, then $\angle C = ?$

  b. If $\angle A = 63°$ and $\angle B = 49°$, find the measure of $\angle C$.

  c. If a triangle has three equal angles, the measure of each angle is ?

  d. If, in $\triangle DEF$, $\angle D = x°$, $\angle E = (x + 30)°$, and $\angle F = (x + 60)°$, find the measure of $\angle F$.

  e. Indicate the maximum number of angles of each type that can be in a triangle.

    (1) Acute   (2) Obtuse   (3) Right   (4) Straight

4. a. Find the hypotenuse of a right triangle if the legs are 6 and 8.

b. Find the leg of a right triangle if the hypotenuse is 13 and one leg is 5.

c. Find the square root of the following:

(1) 169   (2) 144   (3) $\dfrac{1}{4}$

d. Determine whether $\triangle ABC$ is a right triangle if $a = 11$, $b = 12$, and $c = 24$.

5. Classify each of the following quadrilaterals as a parallelogram, a rectangle, a rhombus, a square, a trapezoid, or an isosceles trapezoid. Use the one name that *best* describes the figure.

a.
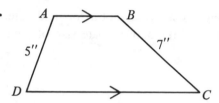

d.
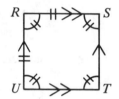

b.
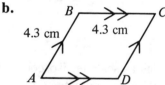

e.

c.
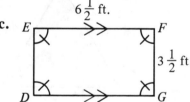

6. Match to the letter *all* the numbers that apply to each of the following geometric figures:

a. Parallelogram

b. Quadrilateral

c. Rhombus

d. Rectangle

e. Square

(1) Equilateral.

(2) Opposite sides of the figure are parallel.

(3) Equiangular.

(4) Four-sided figure.

(5) All of the angles are right angles.

7. Match to the letter *all* the numbers that apply to each of the following geometric figures:

a.	Parallelogram	(1)	Diagonals bisect each other.
b.	Rhombus	(2)	Diagonals are equal.
c.	Square	(3)	Diagonals are perpendicular.
d.	Rectangle	(4)	Diagonals intersect.

8. Complete the chart.

Polygon Name	Number of Sides	Sum of Angles	If the Polygon Is Regular, Find the Size of One Angle.
Triangle			
Quadrilateral			
Pentagon			
Hexagon			
Decagon			

9. Find the perimeter of:

a. A square having a side = 2.6 cm

b. A rectangle with its length = $4\frac{1}{2}''$ and its width = $3\frac{1}{2}''$

c. A triangle with sides of lengths 23, 37, and 49

d. An equilateral triangle whose side = $\frac{2}{3}''$

10. How much fencing is needed to enclose a rectangular garden 6 feet by 4 feet?

11. How far up a building will a 13-foot ladder reach if it is placed 5 feet from the foot of the building?

12. a. If the radius of a wheel of a bike is 14″, find the circumference of the wheel.

b. What is the distance traveled after 300 rotations of the wheel?

Measuring Your Vocabulary

Column II contains the meanings or descriptions of the terms or symbols in Column I, which are used in this chapter.

For each number from Column I, write the letter from Column II that corresponds to the best meaning or description of the term or symbol.

Column I	Column II
1. Triangle	a. A triangle with all three angles equal.
2. Acute angle	b. The side opposite the right angle in a right triangle.
3. Hypotenuse	
4. ∢	c. A triangle with a right angle.
5. Trapezoid	d. A triangle with three acute angles.
6. Rectangle	e. A triangle with one obtuse angle.
7. Circle	f. A four-sided, closed figure whose sides are line segments.
8. Perimeter	
9. Isosceles triangle	g. Two lines in the same plane that never meet.
10. Angle	
11. Degree	h. A quadrilateral whose opposite sides are parallel.
12. Straight angle	
13. Obtuse triangle	i. A line segment joining the center of a circle and any point on the circle.
14. Parallelogram	
15. Rhombus	j. A line segment joining any two points on the circle.
16. Diagonal	
17. °	k. A chord that passes through the center of the circle.
18. Equiangular triangle	l. The distance around a polygon.
19. Obtuse angle	m. The distance around a circle.
20. Vertex	n. The symbol for degree.
21. Parallel lines	o. A quadrilateral with just one pair of opposite sides parallel.
22. Equilateral triangle	
23. Circumference	p. A parallelogram with four equal sides.
24. Quadrilateral	
25. Acute triangle	q. A parallelogram with four equal angles.
26. Chord	
27. Right triangle	r. The symbol for angle.
28. Right angle	s. A line connecting the opposite vertices of a quadrilateral.
29. Diameter	
30. Scalene triangle	t. The set of all points in a plane that are at the same distance from a fixed point called the center.
31. Radius	
	u. A three-sided, closed figure.
	v. A triangle with at least two equal sides.
	w. A triangle with all three sides equal.

Column I	Column II

x. A figure formed by two different rays with a common endpoint.

y. A triangle with no sides equal.

z. The common endpoint of an angle.

aa. The unit of measure of an angle.

bb. An angle that is less than 90°.

cc. An angle that is more than 90° but less than 180°.

dd. An angle that measures 90°.

ee. An angle that measures 180°.

Chapter 5
Managing Money

5-1 *Making Change*

STUDY GUIDE   When we shop we frequently do not have the exact amount of money necessary for the purchase. We pay for the purchase with a $1, $5, $10, or $20 bill and receive *change*.

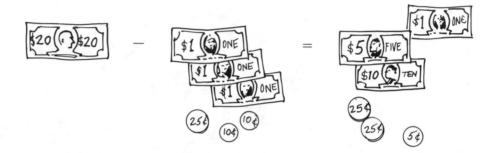

A cashier is a person who accepts the payment and returns change to the purchaser. Many supermarkets and department stores use cash registers that record the purchase price, the amount paid, and the change due the customer. If you purchase items in a store that does not use such a register, you should be able to check the cashier's arithmetic and make certain that you receive the correct change.

FACT FINDING

1. You do not receive change if you pay for a purchase with the

   _____ amount of money.

2. The bills usually offered as payment are _____,

   _____, _____, or _____.

3. The person who accepts money for a purchase and gives you change is

   called a _____.

4. When you receive change you should _____ to see that you receive the correct amount.

MODEL PROBLEM

Brenda purchased a container of milk for 63¢, a pound of butter for $2.49, and a loaf of bread for 79¢. If she pays for the purchases with a $20 bill, how much change will she receive?

Solution: (1) Express the cost of each item as part of a dollar:

$$63¢ = \$0.63$$
$$79¢ = \$0.79$$

(2) Add the three purchases. Remember to line up the decimal points in each amount:

$$
\begin{array}{r}
\$0.63 \\
2.49 \\
+ \ 0.79 \\
\hline
\$3.91
\end{array}
$$

(3) Subtract the total cost of the purchases from $20, writing $20 as a dollar amount:

$$
\begin{array}{rl}
\$20.00 & \text{amount given to cashier} \\
- \ \ 3.91 & \text{cost of purchases} \\
\hline
\$16.09 & \text{amount of change}
\end{array}
$$

EXERCISES

1. Subtract.

a.
$$
\begin{array}{r}
2000 \\
- \ 647 \\
\end{array}
$$

d.
$$
\begin{array}{r}
100 \\
- \ 69 \\
\end{array}
$$

g.
$$
\begin{array}{r}
2000 \\
- \ 1414 \\
\end{array}
$$

j.
$$
\begin{array}{r}
500 \\
- \ 67 \\
\end{array}
$$

m. 1000 − 843

b.
$$
\begin{array}{r}
1000 \\
- \ 817 \\
\end{array}
$$

e.
$$
\begin{array}{r}
1000 \\
- \ 926 \\
\end{array}
$$

h.
$$
\begin{array}{r}
1000 \\
- \ 639 \\
\end{array}
$$

k.
$$
\begin{array}{r}
1500 \\
- \ 436 \\
\end{array}
$$

n. 500 − 269

c.
$$
\begin{array}{r}
500 \\
- \ 241 \\
\end{array}
$$

f.
$$
\begin{array}{r}
100 \\
- \ 87 \\
\end{array}
$$

i.
$$
\begin{array}{r}
2000 \\
- \ 1082 \\
\end{array}
$$

l.
$$
\begin{array}{r}
2500 \\
- \ 2264 \\
\end{array}
$$

o. 2000 − 416

2. Change each amount to a part of a dollar.

a. 13¢   e. 3438¢   i. 4324¢

b. 83¢   f. 7¢   j. 2¢

c. 146¢   g. 50¢   k. 59¢

d. 78¢   h. 168¢   l. 238¢

3. Change each amount to cents.

a.	$1.38	**e.**	$9.07	**i.**	$0.01	**m.**	$1.43
b.	$0.10	**f.**	$0.08	**j.**	$0.69	**n.**	$12.52
c.	$7.50	**g.**	$0.69	**k.**	$17.86	**o.**	$0.07
d.	$13.42	**h.**	$5.10	**l.**	$0.96	**p.**	$35.83

4. Find the amount of change you would receive for each of these purchases:

a.
From a $10 bill	
1 shirt	@ $6.95
1 pair of socks	@ 1.99

f.
From a $5 bill	
1 notebook	@ $2.79
4 pencils	@ 15¢
1 ruler	@ 69¢

b.
From a $5 bill	
1 bread	@ 79¢
1 milk	@ $1.49
1 juice	@ 95¢

g.
From a $10 bill	
2 hamburgers	@ $2.35
1 soda	@ 45¢
1 ice cream	@ 95¢

c.
From a $20 bill	
1 roll of film	@ $3.29
1 flashbulb pack	@ 1.79

h.
From a $20 bill	
1 hammer	@ $14.95
3 scrapers	@ $2
1 wood dough	@ $1.65

d.
From a $10 bill	
1 candy bar	@ 89¢
1 soda	@ 45¢
1 ice cream	@ $1.75

i.
From a $5 bill	
1 newspaper	@ $1.25
2 packs of cigarettes	@ 0.70
4 candy bars	@ 0.30

e.
From a $1 bill	
1 pen	@ 69¢
1 eraser	@ 15¢

j.
From $30	
1 pair of corduroy pants	@ $24.99
Tax of $1.96	

5.
a. How much change did Sara receive from a $20 bill if she made a purchase of $11.37?

b. How much change will Bill return to a customer whose lunch cost $4.10 and who paid with a $20 bill?

c. Maria received a $10 bill in payment for a purchase of $9.27. How much change will Maria return to the customer?

d. How much change will be returned to me if I pay a dinner bill of $26.87 plus a tip of $4.00 with a $50 bill?

e. Gasoline for my car costs $17.63. The attendant tells me that the car needs a quart of oil at a cost of $2.15. How much change will I receive if I pay with a $20 bill?

f. Maria bought 3 cans of frozen orange juice at 69¢ per can and a cake at $1.29. If she pays the cashier with a $5.00 bill, how much change should she receive?

g. Charles picks up a week's cleaning at the dry cleaning shop. He retrieves 2 pairs of slacks at $1.25 each, 3 shirts at 75¢ each, and his raincoat at $3.50. How much change should he receive from a $10 bill?

h. Frank purchases a pair of sneakers at $16.95 and 2 pairs of sweat socks at $2.00 a pair. If Frank must pay $1.47 in sales tax on this purchase and gives the cashier $25.00, how much change should he receive?

EXTRA FOR EXPERTS

1. John made a number of purchases and from a $10 bill offered in payment received change amounting to $3.76. What was the total amount of his purchases?

2. In Exercise 5g, if Charles has only a $5 bill, what items can he retrieve from the cleaner if the shirts are packaged together and the slacks are packaged together?

3. Betsy has $50 birthday money. She purchased shoes at $19.95. With the change she wants to buy a dress at $32.95 or a pair of slacks at $11.95 with a matching blouse at $9.95. Which of these items can she afford?

5-2 Sales Tax

STUDY GUIDE In most cities and states a *sales tax* is added to the price of certain items. The amount of tax varies from place to place, and so does the item taxed. A different rate may be added to food, sporting goods, cosmetics, or restaurant bills.

The sales tax is usually expressed as a percentage of the price of the article being taxed. The tax is calculated by multiplying the amount of the bill by the decimal equivalent of the sales tax percent. For example, the decimal equivalent of 5% is 0.05, of 7% is 0.07, of $8\frac{1}{4}$% is 0.0825, of 10% is 0.10, and of 18% is 0.18. Many stores have a tax chart. The cashier or salesperson checks the chart to determine the sales tax required and adds that amount to the purchase price.

Sales tax = price of items taxed × decimal equivalent of sales tax percent
Purchase price = price of items + tax on these items

FACT FINDING

1. When purchasing certain items a _____ tax is charged.

2. The sales tax is usually expressed as a _____ .

3. To calculate the sales tax, you multiply the _____ by the _____ _____ of the sales tax percent.

4. The sales tax in one city _____ be the same as the tax in
    (may or may not)
another city.

5. The purchase price is _____ than the sales tax.
    (more or less)

MODEL PROBLEM

A can of soda costs $0.50, and an apple costs $0.22.
What is the total cost of the purchase, including sales tax?

Solution: (1)   Find the cost of the items:

$$\begin{array}{r} \$0.50 \\ 0.22 \\ \hline \$0.72 \end{array}$$

(2)   Using the sales tax chart shown below, find the amount of tax corresponding to the total cost of the items:

8% Sales Tax		
$0.01 to $0.10	0¢	$0.06 is the sales tax from the
.11 to .17	1¢	chart for the total purchase price
.18 to .29	2¢	of $0.72.
.30 to .42	3¢	
.43 to .54	4¢	
.55 to .67	5¢	
.68 to .79	6¢	
.80 to .92	7¢	
.93 to 1.06	8¢	

(3)   Find the total purchase price:

$0.72	price of items
+ 0.06	tax
$0.78	total purchase price

From	to	To - Tax
$0.01	to	$0.10 - 0¢
.11	to	.17 - 1¢
.18	to	.29 - 2¢
.30	to	.42 - 3¢
.43	to	.54 - 4¢
.55	to	.67 - 5¢
.68	to	.79 - 6¢
.80	to	.92 - 7¢
.93	to	1.06 - 8¢
1.07	to	1.18 - 9¢
1.19	to	1.31 - 10¢
1.32	to	1.43 - 11¢
1.44	to	1.56 - 12¢
1.57	to	1.68 - 13¢
1.69	to	1.81 - 14¢
1.82	to	1.93 - 15¢
1.94	to	2.06 - 16¢
2.07	to	2.18 - 17¢
2.19	to	2.31 - 18¢
2.32	to	2.43 - 19¢
2.44	to	2.56 - 20¢
2.57	to	2.68 - 21¢
2.69	to	2.81 - 22¢
2.82	to	2.93 - 23¢
2.94	to	3.06 - 24¢
3.07	to	3.18 - 25¢
3.19	to	3.31 - 26¢
3.32	to	3.43 - 27¢
3.44	to	3.56 - 28¢
3.57	to	3.68 - 29¢
3.69	to	3.81 - 30¢
3.82	to	3.93 - 31¢
3.94	to	4.06 - 32¢
4.07	to	4.18 - 33¢
4.19	to	4.31 - 34¢
4.32	to	4.43 - 35¢
4.44	to	4.56 - 36¢
4.57	to	4.68 - 37¢
4.69	to	4.81 - 38¢
4.82	to	4.93 - 39¢
4.94	to	5.06 - 40¢
5.07	to	5.18 - 41¢
5.19	to	5.31 - 42¢
5.32	to	5.43 - 43¢
5.44	to	5.56 - 44¢
5.57	to	5.68 - 45¢
5.69	to	5.81 - 46¢
5.82	to	5.93 - 47¢
5.94	to	6.06 - 48¢
6.07	to	6.18 - 49¢
6.19	to	6.31 - 50¢
6.32	to	6.43 - 51¢
6.44	to	6.56 - 52¢
6.57	to	6.68 - 53¢
6.69	to	6.81 - 54¢
6.82	to	6.93 - 55¢
6.94	to	7.06 - 56¢
7.07	to	7.18 - 57¢
7.19	to	7.31 - 58¢
7.32	to	7.43 - 59¢
7.44	to	7.56 - 60¢
7.57	to	7.68 - 61¢
7.69	to	7.81 - 62¢
7.82	to	7.93 - 63¢
7.94	to	8.06 - 64¢
8.07	to	8.18 - 65¢
8.19	to	8.31 - 66¢
8.32	to	8.43 - 67¢
8.44	to	8.56 - 68¢
8.57	to	8.68 - 69¢
8.69	to	8.81 - 70¢
8.82	to	8.93 - 71¢
8.94	to	9.06 - 72¢
9.07	to	9.18 - 73¢
9.19	to	9.31 - 74¢
9.32	to	9.43 - 75¢
9.44	to	9.56 - 76¢
9.57	to	9.68 - 77¢
9.69	to	9.81 - 78¢
9.82	to	9.93 - 79¢
9.94	to	10.06 - 80¢
10.07	to	10.18 - 81¢
10.19	to	10.31 - 82¢
10.32	to	10.43 - 83¢
10.44	to	10.56 - 84¢
10.57	to	10.68 - 85¢
10.69	to	10.81 - 86¢
10.82	to	10.93 - 87¢
10.94	to	11.06 - 88¢
11.07	to	11.18 - 89¢
11.19	to	11.31 - 90¢
11.32	to	11.43 - 91¢
11.44	to	11.56 - 92¢
11.57	to	11.68 - 93¢
11.69	to	11.81 - 94¢
11.82	to	11.93 - 95¢
11.94	to	12.06 - 96¢
12.07	to	12.18 - 97¢
12.19	to	12.31 - 98¢
12.32	to	12.43 - 99¢
12.44	to	12.56 - 1.00
12.57	to	12.68 - 1.01
12.69	to	12.81 - 1.02
12.82	to	12.93 - 1.03
12.94	to	13.06 - 1.04
13.07	to	13.18 - 1.05
13.19	to	13.31 - 1.06
13.32	to	13.43 - 1.07
13.44	to	13.56 - 1.08
13.57	to	13.68 - 1.09
13.69	to	13.81 - 1.10
13.82	to	13.93 - 1.11
13.94	to	14.06 - 1.12
14.07	to	14.18 - 1.13
14.19	to	14.31 - 1.14
14.32	to	14.43 - 1.15
14.44	to	14.56 - 1.16
14.57	to	14.68 - 1.17
14.69	to	14.81 - 1.18
14.82	to	14.93 - 1.19
14.94	to	15.06 - 1.20
15.07	to	15.18 - 1.21
15.19	to	15.31 - 1.22
15.32	to	15.43 - 1.23
15.44	to	15.56 - 1.24
15.57	to	15.68 - 1.25
15.69	to	15.81 - 1.26
15.82	to	15.93 - 1.27
15.94	to	16.06 - 1.28
16.07	to	16.18 - 1.29
16.19	to	16.31 - 1.30
16.32	to	16.43 - 1.31
16.44	to	16.56 - 1.32
16.57	to	16.68 - 1.33
16.69	to	16.81 - 1.34
16.82	to	16.93 - 1.35
16.94	to	17.06 - 1.36
17.07	to	17.18 - 1.37
17.19	to	17.31 - 1.38
17.32	to	17.43 - 1.39
17.44	to	17.56 - 1.40
17.57	to	17.68 - 1.41
17.69	to	17.81 - 1.42
17.82	to	17.93 - 1.43
17.94	to	18.06 - 1.44
18.07	to	18.18 - 1.45
18.19	to	18.31 - 1.46
18.32	to	18.43 - 1.47
18.44	to	18.56 - 1.48
18.57	to	18.68 - 1.49
18.69	to	18.81 - 1.50
18.82	to	18.93 - 1.51
18.94	to	19.06 - 1.52
19.07	to	19.18 - 1.53
19.19	to	19.31 - 1.54
19.32	to	19.43 - 1.55
19.44	to	19.56 - 1.56
19.57	to	19.68 - 1.57
19.69	to	19.81 - 1.58
19.82	to	19.93 - 1.59
19.94	to	20.06 - 1.60
20.07	to	20.18 - 1.61
20.19	to	20.31 - 1.62
20.32	to	20.43 - 1.63
20.44	to	20.56 - 1.64
20.57	to	20.68 - 1.65
20.69	to	20.81 - 1.66
20.82	to	20.93 - 1.67
20.94	to	21.06 - 1.68
21.07	to	21.18 - 1.69
21.19	to	21.31 - 1.70
21.32	to	21.43 - 1.71
21.44	to	21.56 - 1.72
21.57	to	21.68 - 1.73
21.69	to	21.81 - 1.74
21.82	to	21.93 - 1.75
21.94	to	22.06 - 1.76
22.07	to	22.18 - 1.77
22.19	to	22.31 - 1.78
22.32	to	22.43 - 1.79
22.44	to	22.56 - 1.80
22.57	to	22.68 - 1.81
22.69	to	22.81 - 1.82
22.82	to	22.93 - 1.83
22.94	to	23.06 - 1.84
23.07	to	23.18 - 1.85
23.19	to	23.31 - 1.86
23.32	to	23.43 - 1.87
23.44	to	23.56 - 1.88
23.57	to	23.68 - 1.89
23.69	to	23.81 - 1.90
23.82	to	23.93 - 1.91
23.94	to	24.06 - 1.92
24.07	to	24.18 - 1.93
24.19	to	24.31 - 1.94
24.32	to	24.43 - 1.95
24.44	to	24.56 - 1.96
24.57	to	24.68 - 1.97
24.69	to	24.81 - 1.98
24.82	to	24.93 - 1.99
24.94	to	25.06 - 2.00
25.07	to	25.18 - 2.01
25.19	to	25.31 - 2.02
25.32	to	25.43 - 2.03
25.44	to	25.56 - 2.04
25.57	to	25.68 - 2.05
25.69	to	25.81 - 2.06
25.82	to	25.93 - 2.07
25.94	to	26.06 - 2.08
26.07	to	26.18 - 2.09
26.19	to	26.31 - 2.10
26.32	to	26.43 - 2.11
26.44	to	26.56 - 2.12
26.57	to	26.68 - 2.13
26.69	to	26.81 - 2.14
26.82	to	26.93 - 2.15
26.94	to	27.06 - 2.16
27.07	to	27.18 - 2.17
27.19	to	27.31 - 2.18
27.32	to	27.43 - 2.19
27.44	to	27.56 - 2.20
27.57	to	27.68 - 2.21
27.69	to	27.81 - 2.22
27.82	to	27.93 - 2.23
27.94	to	28.06 - 2.24
28.07	to	28.18 - 2.25
28.19	to	28.31 - 2.26
28.32	to	28.43 - 2.27
28.44	to	28.56 - 2.28
28.57	to	28.68 - 2.29
28.69	to	28.81 - 2.30
28.82	to	28.93 - 2.31
28.94	to	29.06 - 2.32
29.07	to	29.18 - 2.33
29.19	to	29.31 - 2.34
29.32	to	29.43 - 2.35
29.44	to	29.56 - 2.36
29.57	to	29.68 - 2.37
29.69	to	29.81 - 2.38
29.82	to	29.93 - 2.39
29.94	to	30.06 - 2.40
30.07	to	30.18 - 2.41
30.19	to	30.31 - 2.42
30.32	to	30.43 - 2.43
30.44	to	30.56 - 2.44
30.57	to	30.68 - 2.45
30.69	to	30.81 - 2.46
30.82	to	30.93 - 2.47
30.94	to	31.06 - 2.48
31.07	to	31.18 - 2.49
31.19	to	31.31 - 2.50
31.32	to	31.43 - 2.51
31.44	to	31.56 - 2.52
31.57	to	31.68 - 2.53
31.69	to	31.81 - 2.54
31.82	to	31.93 - 2.55
31.94	to	32.06 - 2.56
32.07	to	32.18 - 2.57
32.19	to	32.31 - 2.58
32.32	to	32.43 - 2.59
32.44	to	32.56 - 2.60
32.57	to	32.68 - 2.61
32.69	to	32.81 - 2.62
32.82	to	32.93 - 2.63
32.94	to	33.06 - 2.64
33.07	to	33.18 - 2.65
33.19	to	33.31 - 2.66
33.32	to	33.43 - 2.67
33.44	to	33.56 - 2.68
33.57	to	33.68 - 2.69
33.69	to	33.81 - 2.70
33.82	to	33.93 - 2.71
33.94	to	34.06 - 2.72
34.07	to	34.18 - 2.73
34.19	to	34.31 - 2.74
34.32	to	34.43 - 2.75
34.44	to	34.56 - 2.76
34.57	to	34.68 - 2.77
34.69	to	34.81 - 2.78
34.82	to	34.93 - 2.79
34.94	to	35.06 - 2.80
35.07	to	35.18 - 2.81
35.19	to	35.31 - 2.82
35.32	to	35.43 - 2.83
35.44	to	35.56 - 2.84
35.57	to	35.68 - 2.85
35.69	to	35.81 - 2.86
35.82	to	35.93 - 2.87
35.94	to	36.06 - 2.88
36.07	to	36.18 - 2.89
36.19	to	36.31 - 2.90
36.32	to	36.43 - 2.91
36.44	to	36.56 - 2.92
36.57	to	36.68 - 2.93
36.69	to	36.81 - 2.94
36.82	to	36.93 - 2.95
36.94	to	37.06 - 2.96
37.07	to	37.18 - 2.97
37.19	to	37.31 - 2.98
37.32	to	37.43 - 2.99
37.44	to	37.56 - 3.00
37.57	to	37.68 - 3.01
37.69	to	37.81 - 3.02
37.82	to	37.93 - 3.03
37.94	to	38.06 - 3.04
38.07	to	38.18 - 3.05
38.19	to	38.31 - 3.06
38.32	to	38.43 - 3.07
38.44	to	38.56 - 3.08
38.57	to	38.68 - 3.09
38.69	to	38.81 - 3.10
38.82	to	38.93 - 3.11
38.94	to	39.06 - 3.12
39.07	to	39.18 - 3.13
39.19	to	39.31 - 3.14
39.32	to	39.43 - 3.15
39.44	to	39.56 - 3.16
39.57	to	39.68 - 3.17
39.69	to	39.81 - 3.18
39.82	to	39.93 - 3.19
39.94	to	40.06 - 3.20
40.07	to	40.18 - 3.21
40.19	to	40.31 - 3.22
40.32	to	40.43 - 3.23
40.44	to	40.56 - 3.24
40.57	to	40.68 - 3.25
40.69	to	40.81 - 3.26
40.82	to	40.93 - 3.27
40.94	to	41.06 - 3.28
41.07	to	41.18 - 3.29
41.19	to	41.31 - 3.30
41.32	to	41.43 - 3.31
41.44	to	41.56 - 3.32
41.57	to	41.68 - 3.33
41.69	to	41.81 - 3.34
41.82	to	41.93 - 3.35
41.94	to	42.06 - 3.36

On higher amounts, compute the tax by multiplying the amount of sale by 8% and rounding the result to the nearest whole cent.

Use the 8% sales tax chart on page 237 for Exercise 1.

1. Indicate the amount of tax you would pay on the following purchases:

   a. $0.46   **d.** $0.95   **g.** $0.25

   b. $0.73   **e.** $0.87   **h.** $0.62

   c. $0.15   **f.** $0.35   **i.** $0.51

2. Indicate the amount of tax you would pay on the following purchases if they are taxed at a 9% rate. (*Reminder:* 9% is equivalent to 0.09.)

   a. $14.12   **d.** $4.36   **g.** $35.04

   b. $23.19   **e.** $10.89   **h.** $6.72

   c. $43.57   **f.** $29.99   **i.** $40.90

3. Donna bought a 5-foot toboggan for $22.99, a 36-inch snoboggan for $9.99, and a sled for $12.99. She paid an 8% sales tax on her purchases.

   a. What was the total amount of her purchases?

   b. What was the sales tax?

   c. What was the total cost?

   d. If she paid 6% sales tax on her purchases, find the sales tax, the total cost, and the amount of money she saved at this lower tax rate.

4. Willy purchased two knit shirts for $2.99 each, a hooded sweatshirt for $7.99, and three long-sleeved velour shirts at $7.44 each.

   a. What was the total amount of his purchases?

   b. Find the sales tax if the rate is 8%.

   c. What was the total cost?

   d. Find the sales tax if the rate was 9%. What was the total cost at this rate?

In a specific county the sales tax is 7%. Use the 7% sales tax chart on page 239 for Exercises 5 and 9.

5. In a department store in this county you purchase two sheets at $4.99 each and four pillowcases at $1.99 each.

   a. What was the total amount of your purchases?

   b. Using the chart, find the sales tax if the rate is 7%.

   c. What was the total cost?

   d. Find the sales tax if the rate is 9%. If the rate is increased to 9%, by how much is the total cost increased?

7% SALES TAX

Combined 4% New York State Tax and 3% Local Tax
Effective June 1, 1971

Amount		Tax	Amount		Tax	Amount		Tax	Amount		Tax
$ .01	to	$ .10 - 0¢									
.11	to	.20 - 1¢	7.08	to	7.21 - 50¢				20.08	to	20.21 - 1.41
.21	to	.33 - 2¢	7.22	to	7.35 - 51¢				20.22	to	20.35 - 1.42
.34	to	.47 - 3¢	7.36	to	7.49 - 52¢				20.36	to	20.49 - 1.43
.48	to	.62 - 4¢	7.50	to	7.64 - 53¢				20.50	to	20.64 - 1.44
.63	to	.76 - 5¢	7.65	to	7.78 - 54¢				20.65	to	20.78 - 1.45
.77	to	.91 - 6¢	7.79	to	7.92 - 55¢				20.79	to	20.92 - 1.46
.92	to	1.07 - 7¢	7.93	to	8.07 - 56¢				20.93	to	21.07 - 1.47
1.08	to	1.21 - 8¢	8.08	to	8.21 - 57¢	14.08	to	14.21 - 99¢	21.08	to	21.21 - 1.48
1.22	to	1.35 - 9¢	8.22	to	8.35 - 58¢	14.22	to	14.35 - 1.00	21.22	to	21.35 - 1.49
1.36	to	1.49 - 10¢	8.36	to	8.49 - 59¢	14.36	to	14.49 - 1.01	21.36	to	21.49 - 1.50
1.50	to	1.64 - 11¢	8.50	to	8.64 - 60¢	14.50	to	14.64 - 1.02	21.50	to	21.64 - 1.51
1.65	to	1.78 - 12¢	8.65	to	8.78 - 61¢	14.65	to	14.78 - 1.03	21.65	to	21.78 - 1.52
1.79	to	1.92 - 13¢	8.79	to	8.92 - 62¢	14.79	to	14.92 - 1.04	21.79	to	21.92 - 1.53
1.93	to	2.07 - 14¢	8.93	to	9.07 - 63¢	14.93	to	15.07 - 1.05	21.93	to	22.07 - 1.54
2.08	to	2.21 - 15¢	9.08	to	9.21 - 64¢	15.08	to	15.21 - 1.06	22.08	to	22.21 - 1.55
2.22	to	2.35 - 16¢	9.22	to	9.35 - 65¢	15.22	to	15.35 - 1.07	22.22	to	22.35 - 1.56
2.36	to	2.49 - 17¢	9.36	to	9.49 - 66¢	15.36	to	15.49 - 1.08	22.36	to	22.49 - 1.57
2.50	to	2.64 - 18¢	9.50	to	9.64 - 67¢	15.50	to	15.64 - 1.09	22.50	to	22.64 - 1.58
2.65	to	2.78 - 19¢	9.65	to	9.78 - 68¢	15.65	to	15.78 - 1.10	22.65	to	22.78 - 1.59
2.79	to	2.92 - 20¢	9.79	to	9.92 - 69¢	15.79	to	15.92 - 1.11	22.79	to	22.92 - 1.60
2.93	to	3.07 - 21¢	9.93	to	10.07 - 70¢	15.93	to	16.07 - 1.12	22.93	to	23.07 - 1.61
3.08	to	3.21 - 22¢	10.08	to	10.21 - 71¢	16.08	to	16.21 - 1.13	23.08	to	23.21 - 1.62
3.22	to	3.35 - 23¢	10.22	to	10.35 - 72¢	16.22	to	16.35 - 1.14	23.22	to	23.35 - 1.63
3.36	to	3.49 - 24¢	10.36	to	10.49 - 73¢	16.36	to	16.49 - 1.15	23.36	to	23.49 - 1.64
3.50	to	3.64 - 25¢	10.50	to	10.64 - 74¢	16.50	to	16.64 - 1.16	23.50	to	23.64 - 1.65
3.65	to	3.78 - 26¢	10.65	to	10.78 - 75¢	16.65	to	16.78 - 1.17	23.65	to	23.78 - 1.66
3.79	to	3.92 - 27¢	10.79	to	10.92 - 76¢	16.79	to	16.92 - 1.18	23.79	to	23.92 - 1.67
3.93	to	4.07 - 28¢	10.93	to	11.07 - 77¢	16.93	to	17.07 - 1.19	23.93	to	24.07 - 1.68
4.08	to	4.21 - 29¢	11.08	to	11.21 - 78¢	17.08	to	17.21 - 1.20	24.08	to	24.21 - 1.69
4.22	to	4.35 - 30¢	11.22	to	11.35 - 79¢	17.22	to	17.35 - 1.21	24.22	to	24.35 - 1.70
4.36	to	4.49 - 31¢	11.36	to	11.49 - 80¢	17.36	to	17.49 - 1.22	24.36	to	24.49 - 1.71
4.50	to	4.64 - 32¢	11.50	to	11.64 - 81¢	17.50	to	17.64 - 1.23	24.50	to	24.64 - 1.72
4.65	to	4.78 - 33¢	11.65	to	11.78 - 82¢	17.65	to	17.78 - 1.24	24.65	to	24.78 - 1.73
4.79	to	4.92 - 34¢	11.79	to	11.92 - 83¢	17.79	to	17.92 - 1.25	24.79	to	24.92 - 1.74
4.93	to	5.07 - 35¢	11.93	to	12.07 - 84¢	17.93	to	18.07 - 1.26	24.93	to	25.07 - 1.75
5.08	to	5.21 - 36¢	12.08	to	12.21 - 85¢	18.08	to	18.21 - 1.27	25.08	to	25.21 - 1.76
5.22	to	5.35 - 37¢	12.22	to	12.35 - 86¢	18.22	to	18.35 - 1.28	25.22	to	25.35 - 1.77
5.36	to	5.49 - 38¢	12.36	to	12.49 - 87¢	18.36	to	18.49 - 1.29	25.36	to	25.49 - 1.78
5.50	to	5.64 - 39¢	12.50	to	12.64 - 88¢	18.50	to	18.64 - 1.30	25.50	to	25.64 - 1.79
5.65	to	5.78 - 40¢	12.65	to	12.78 - 89¢	18.65	to	18.78 - 1.31	25.65	to	25.78 - 1.80
5.79	to	5.92 - 41¢	12.79	to	12.92 - 90¢	18.79	to	18.92 - 1.32	25.79	to	25.92 - 1.81
5.93	to	6.07 - 42¢	12.93	to	13.07 - 91¢	18.93	to	19.07 - 1.33	25.93	to	26.07 - 1.82
6.08	to	6.21 - 43¢	13.08	to	13.21 - 92¢	19.08	to	19.21 - 1.34	26.08	to	26.21 - 1.83
6.22	to	6.35 - 44¢	13.22	to	13.35 - 93¢	19.22	to	19.35 - 1.35	26.22	to	26.35 - 1.84
6.36	to	6.49 - 45¢	13.36	to	13.49 - 94¢	19.36	to	19.49 - 1.36	26.36	to	26.49 - 1.85
6.50	to	6.64 - 46¢	13.50	to	13.64 - 95¢	19.50	to	19.64 - 1.37	26.50	to	26.64 - 1.86
6.65	to	6.78 - 47¢	13.65	to	13.78 - 96¢	19.65	to	19.78 - 1.38	26.65	to	26.78 - 1.87
6.79	to	6.92 - 48¢	13.79	to	13.92 - 97¢	19.79	to	19.92 - 1.39	26.79	to	26.92 - 1.88
6.93	to	7.07 - 49¢	13.93	to	14.07 - 98¢	19.93	to	20.07 - 1.40	26.93	to	27.07 - 1.89

On higher amounts, compute the tax by multiplying the amount of sale by 7% and rounding the result to the nearest whole cent.

6. In a county with a 5% sales tax rate, Janine purchased a jacket at $41.95. How much tax must she pay on the item?

7. In a county with a 4% sales tax rate, Michael purchased two pairs of shoes on sale at $18.95 a pair. He must pay taxes on the shoes. What was the total cost of his purchases?

8. Jane bought a skirt at $14.95 and a blouse at $8.95 in a store with a 6% sales tax rate. What was the total cost of her purchases?

9. At a food market in Arbor County Frank purchased 3 liters of soda at 69¢ each. Soda is taxed at 7 percent. How much change did Frank receive if he paid for the purchase with a $5.00 bill?

10. Mary bought a six-pack of 7-Up at $1.99 and a box of candy bars at $5.95. If both items are taxable, what would Mary's total bill be in a city with $8\frac{1}{4}$% sales tax? (*Reminder:* $8\frac{1}{4}$% is equivalent to 0.0825.) What would her total bill be in a state with a 4% sales tax rate?

EXTRA FOR EXPERTS

1. Maria purchased 3 sandwiches at $1.75 each, 2 cans of soda at $0.50 each, 1 coffee at $0.45, and a Danish at $0.60. If an 8% sales tax is paid, what is the total cost of the purchases? How much change should Maria receive from a $20 bill? If a 10% sales tax is paid, what is the difference in the change that Maria would receive?

2. In a sports shop in a county with a 6% sales tax, Oliver purchased a basketball for $5.95, 2 baseballs at $1.39 each, and an aluminum baseball bat at $19.95. Find the total cost, including tax. Oliver put down a deposit of $10. What is his balance? What change should he receive if he pays the balance with a $20 bill?

3. A suit costs $95 at a department store. In New York City an $8\frac{1}{4}$% sales tax must be paid. In New Jersey there is no sales tax on clothing. If it costs Wilton $4.50 round-trip on the bus to go to a New Jersey shopping mall, will he save money if he goes there to buy the suit? Would he always save money if he went to New Jersey to buy his clothes? Give an example to explain your answer.

5-3.1 *Comparing Costs*

STUDY GUIDE Most shoppers compare prices. Supermarkets are compelled by law to display the unit price of each item they sell. The *unit price* is the cost of an item per unit of measure. It may be expressed, for example, as price per quart, price per pound, or price per gallon.

This practice gives the consumer a chance to choose the "best" or "better" buy. When the unit prices for two articles are known and the units are the same (for example, the price per quart), the lower cost is considered the "better" buy.

FACT FINDING

1. The unit price of an item is its price per unit of _____ .

2. Unit price is expressed as price per _____ or price per

_____ .

3. Finding the "better" buy means comparing the _____ prices for the same unit of measure.

4. The lower unit cost for the same item is considered the "_____" buy.

MODEL PROBLEMS

1. Which is the better buy: a half-gallon bottle of Brand C soda selling for $0.99 or a gallon of Brand P soda selling for $1.59?

   Solution: By checking the unit price shown on the shelf below each item we find:

Unit Price	Retail Price		Unit Price	Retail Price
$0.495 per qt.	$0.99 per ½ gal.		$0.3975 per qt.	$1.59 per gal.
Brand C			Brand P	

By looking under "Unit Price," we see that Brand P is cheaper because it is $0.3975 per quart as compared with Brand C, which is $0.495 per quart.

2. The unit price for each of three brands would look as shown below on the supermarket shelf. Which is the "best" buy?

Unit Price	Retail Price
$5.20 per quart	$0.65 per 4 oz.

Brand X

or

Unit Price	Retail Price
$5.16 per quart	$1.29 per half–pint

Brand Y

or

Unit Price	Retail Price
$4.98 per quart	$2.49 per pint

Brand Z

Solution: Checking the unit price for each brand, you find that the prices are all for the same unit of measure, the quart, and thus the lowest unit price is considered the "best" buy. Therefore, Brand Z is the "best" buy.

EXERCISES

In each of these indicate the "better" buy:

a.

Unit Price	Retail Price
$1.52 per qt.	$0.19 4. oz.

Brand A

or

Unit Price	Retail Price
$1.56 per qt.	39¢ 8 oz.

Brand B

b.

Unit Price	Retail Price
$4.495 per qt.	59¢ 4.2 oz.

Brand C

or

Unit Price	Retail Price
$4.60 per qt.	$0.69 0.3 pt.

Brand D

c.

Unit Price	Retail Price
72¢ per pt.	9¢ 2 oz.

Brand E

or

Unit Price	Retail Price
$0.80 per pt.	$0.25 5 oz.

Brand F

d.

Unit Price	Retail Price
$0.795 per pt.	$1.59 1 qt.

Brand G

or

Unit Price	Retail Price
$0.65 per pt.	$1.95 1½ qt.

Brand H

e.

Unit Price	Retail Price
6.3¢ per oz.	19¢ 3 oz.

Brand I

or

Unit Price	Retail Price
6.875¢ per oz.	55¢ 8 oz.

Brand J

f.

Unit Price	Retail Price
$3.60 per gal.	45¢ per pt.

Brand K

or

Unit Price	Retail Price
$3.56 per gal.	89¢ per qt.

Brand L

g.

Unit Price	Retail Price
80¢ per pt.	5¢ per oz.

Brand M

or

Unit Price	Retail Price
76¢ per pt.	19¢ 4 oz.

Brand N

h.

Unit Price	Retail Price
$0.49 per oz.	$1.57 3.2 oz.

Brand O

or

Unit Price	Retail Price
58¢ per oz.	29¢ 8 oz.

Brand P

i.

Unit Price	Retail Price
$0.90 per qt.	45¢ per pt.

Brand Q

or

Unit Price	Retail Price
92¢ per qt.	23¢ 8 oz.

Brand R

j.

Unit Price	Retail Price
$2.983 per pt.	$1.79 0.6 pt.

Brand S

or

Unit Price	Retail Price
$2.991 per pt.	$3.59 1.2 pt.

Brand T

5-3.2 *Comparing Costs (Continued)*

Very often we find that the units of measure are different and conversion is necessary in order to make a comparison and to determine the "best" buy.

MODEL PROBLEM

Three brands of olive oil are sold at the supermarket. Brand X is sold in 4-ounce cans and costs 65¢; Brand Y is sold in $\frac{1}{2}$-pint cans and costs $1.29; and Brand Z is sold in pint cans and costs $2.49. Which of the three brands is considered the "best" buy?

Liquid Measure	
1 pint (pt.)	= 16 ounces (oz.)
1 quart (qt.)	= 2 pt.
1 gallon (gal.)	= 4 qt.

Solution: Compare the three brands, X, Y, and Z, in tabular form.

Brand	4 oz.	$\frac{1}{2}$ pt.	1 pt.	1 qt.
X	$0.65			8 × $0.65 = $5.20
Y		$1.29		4 × $1.29 = $5.16
Z			$2.49	2 × $2.49 = $4.98

By changing all of the prices to price per quart, we determine the "best" buy. Thus, if price is the only consideration, Brand Z is the "best" buy.

EXERCISES

1. Change each liquid measure to its equivalent.

    a. 8 oz. = __ pt.

    b. 2 qt. = __ pt.

    c. 1 gal. = __ pt.

    d. 1 qt. = __ oz.

    e. 3 gal. = __ qt.

    f. __ oz. = 1 gal.

    g. 12 oz. = __ pt.

    h. __ qt. = 2 gal.

    i. __ gal. = 4 qt.

    j. 1 qt. 1 pt. = __ oz.

In Exercises 2–6, determine the "better" buy:

2. Brand X Tuna   Brand Y Tuna
4 ounces $1.65   16 ounces $2.69

3. Brand R Orange Juice   Brand P Orange Juice
$1.23 per quart $0.39 per 4 ounce

4. Brand D Liquid Cleaner   Brand I Liquid Cleaner
$0.69 per $\frac{1}{2}$ pint $4.29 per gallon

5. Brand C Liquid Bleach   Brand R Liquid Bleach
$0.99 per gallon $0.34 per quart

6. Brand L Milk   Brand T Milk
$0.75 per quart $3.54 per gallon

7. Marsha is buying whole wheat bread. The 12-ounce package of Brand A is 73¢. The 16-ounce package of Brand B is 95¢. Which is the "better" buy?

8. Mouthwash is $1.29 for the 16-ounce size. If the gallon size is marked $6.95, which is the "better" buy?

9. Manuel is purchasing orange juice. A quart costs 93¢, and a half-gallon costs $1.72. Which is the "better" buy? Is the larger size of an item always the "better" buy?

10. Francis wants soda with his dinner. The supermarket is having a sale on 12-ounce cans of soda. If a six-pack costs $1.99 on sale and a 32-ounce bottle costs 83¢, which is the "better" buy?

EXTRA FOR EXPERTS

1. Jackson has a 50¢ discount ticket on any size jar of a certain brand of instant coffee. If the 12-ounce size is $2.95 and the $\frac{1}{2}$-pound (16-ounce) size is $3.95, which is the better purchase? What price will Jackson have to pay if he uses his coupon and purchases the "better" buy?

2. In Extra 1, if the sales tax is 8% on coffee and Jackson pays for his "better" buy with a $5 bill, what change will he receive?

3. José has a 75¢ coupon good on a gallon or more of milk. A quart is 69¢. A gallon is $2.27. The shelf life of milk is 5 days. If José drinks a pint of milk a day, should he purchase the gallon of milk? Explain your answer.

5-4.1 *Computing the Retail Price*

The unit price is the cost of an item per unit of measure, but the item you wish to purchase may be sold in small, medium, and large sizes. To find the *retail price* of an item, you must do the following:

1. Make each measure the same.
2. Multiply the unit price by the measure.

Retail price = unit price × measure

or

RP = UP × M.

FACT FINDING

1. Items you wish to purchase may be sold in _____ or _____ sizes than the measure of the unit price.

2. Before you can calculate the retail price, you must make both _____ the same.

3. To find the retail price you _____ the unit price by the measure.

MODEL PROBLEM

As a part-time worker in a supermarket, you must compute the retail price of items from the unit price. If the unit price of a pound of string beans is \$1.20 per pound, what will 4 ounces cost?

Solution: Express 4 ounces as part of a pound.

Dry Measure
1 pound (lb.) = 16 oz.

Since 16 oz. = 1 lb.,

$$\frac{4\ oz.}{1\ lb.} = \frac{4\ oz.}{16\ oz.} = \frac{1}{4}\ lb. = \frac{1}{4} \times \frac{16}{1} = 4\ oz.$$

Retail price = unit price × measure.

$$RP = UP \times M$$

$$= \$1.20 \times \frac{1}{4}$$

$$= \$\overset{.30}{\cancel{1.20}} \times \frac{1}{\underset{1}{\cancel{4}}}$$

$$RP = \$0.30.$$

1. Change each measure to its equivalent.

   a. 8 oz. = __ lb.   f. 2 lb. = __ oz.

   b. 1 pt. = __ qt.   g. 12 oz. = __ pt.

   c. 3 qt. = __ gal.   h. 32 oz. = __ lb.

   d. 2 gal. = __ pt.   i. 1 pt. = __ gal.

   e. 24 oz. = __ lb.   j. 2 oz. = __ lb.

2. Compute the retail price of each item.

   Remember: Retail price = unit price × measure.

a.

Unit Price	Retail Price
$2.40 per lb.	? 1½ lb.

f.

Unit Price	Retail Price
$0.78 per qt.	? 1 pt.

b.

Unit Price	Retail Price
$0.18 per oz.	? 1 qt.

g.

Unit Price	Retail Price
$0.695 per 6 oz.	? 12 oz.

c.

Unit Price	Retail Price
$3.16 per gal.	? 1 qt.

h.

Unit Price	Retail Price
$1.84 per pt.	? 6 oz.

d.

Unit Price	Retail Price
$0.90 per pt.	? 12 oz.

i.

Unit Price	Retail Price
$1.38 per lb.	? 8 oz.

e.

Unit Price	Retail Price
$1.99 per lb.	? 3 lb.

j.

Unit Price	Retail Price
$0.595 per pt.	? 1 qt.

Many items in the meat and the fruit and vegetable departments cannot be measured in exact units and may be purchased in packages weighing, for example, 0.8 or 0.62 lb. What is the retail price of 0.87 lb. of broccoli if its unit price is $0.79 per lb.?

Unit Price	Retail Price
$0.79 per lb.	? 0.87 lb.

Solution: Since the units of measure are alike, multiply the unit price by the measure of the item:

$$\begin{array}{rl} \$0.79 & \text{unit price} \\ \times\ 0.87 & \text{lb. weight} \\ \hline 553 & \\ 632 & \\ \hline \$0.6873 & \end{array}$$

RP = UP × M
RP = $0.79 × 0.87
RP = $0.69

Since there are *two* decimal places in 0.79 and *two* in 0.87, there is a total of *four* decimal places in the product, 0.6873. Rounding off $0.6873 to the nearest cent results in $0.69.

Since the smallest unit of money we will use to pay for an item is cents, we should round off the price to two decimal places as follows: 0.4739. The third decimal place, 3, is less than 5. Therefore this amount is 0.47 to the nearest cent.

0.585 = 0.59 to the nearest cent because 5 is in the third decimal place.
Similarly, 0.6123 is 0.61 to the nearest cent;
0.7468 is 0.75 to the nearest cent.

1. Round off to the nearest cent:

a. 0.6326		**d.** 0.5719		**g.** 0.2771		**j.** 0.1863
b. 0.5681		**e.** 0.3274		**h.** 0.8241		**k.** 0.2541
c. 0.784		**f.** 0.4197		**i.** 0.9146		**l.** 0.7549

2. Compute the retail price to the nearest cent.

a.

Unit Price	Retail Price
$1.49 per lb.	? 1.2 lb.

f.

Unit Price	Retail Price
$2.59 per lb.	? 2.13 lb.

b.

Unit Price	Retail Price
$0.89 per lb.	? 0.34 lb.

g.

Unit Price	Retail Price
$0.39 per oz.	? 3.41 oz.

c.

Unit Price	Retail Price
$1.19 per lb.	? 3.2 lb.

h.

Unit Price	Retail Price
$1.29 per lb.	? 0.64 lb.

d.

Unit Price	Retail Price
$0.19 per oz.	? 2.3 oz.

i.

Unit Price	Retail Price
$0.29 per oz.	? 4.03 oz.

e.

Unit Price	Retail Price
$2.99 per lb.	? 0.94 lb.

j.

Unit Price	Retail Price
$0.69 per oz.	? 0.28 oz.

3. Honey sells for $1.59 per pound. Maria needs 12 ounces for a honey carrot loaf. How much will 12 ounces cost?

4. Chopped meat for hamburgers sells for $1.89 a pound. Juan is having a barbeque for 10 people and needs $\frac{1}{2}$ pound per person. How many pounds does he need for the 10 people? What is the total amount that Juan spends?

5. Vinegar sells for $1.09 a quart. A salad dressing recipe calls for 1 pint. How much will the pint of vinegar cost?

6. Canned pears are listed as $0.49 a pound. The price on the 20-ounce can is missing. Compute the price.

7. Melody needs 12 ounces of tuna fish for a casserole. The delicatessen has only 7-ounce cans for $1.69. How much will the tuna cost? Will Melody have any left over?

8. An 8-ounce box of tea bags sells for $2.69. The box will make 172 ounces of liquid tea. A liquid concentrate of tea costs $3.16 for 32 ounces, and will make 256 ounces of liquid tea. Which is the "better" buy, the box or the bottle?

1. Mrs. Brown ordered the following items from the butcher:

   $\frac{1}{2}$ pound of beef at $2.89 a pound

   8 ounces of chicken at $1.69 a pound

   $2\frac{1}{2}$ pounds of veal at $4.16 a pound

What is the total cost of her meat bill?

2. In the vegetable store Mrs. Brown bought:

   1 head of broccoli at $0.69

   4.2 pounds of potatoes at $0.39 a pound

   2.3 pounds of onions at $0.33 a pound

   12 ounces of mushrooms at $1.99 a pound

What was the amount spent in the vegetable store? What should have been the change from a $10 bill?

5-4.2 *Computing the Unit Price*

STUDY GUIDE   The *unit price* of an item can be determined if the retail price is known as well as the measure. To find the unit price you divide the retail price by the measure.

The measure must always be written in terms of the measure used in the unit price. If the unit price measure is in terms of pounds and ounces are given, the ounces must first be written as a fractional part of a pound.

$$\text{Unit price} = \frac{\text{retail price}}{\text{measure}}$$

or

$$UP = \frac{RP}{M}.$$

FACT FINDING

1. If the measure and the retail price of an item are known, the

   _____ _____ can be determined.

2. Dividing the retail price by its measure will result in the _____ price.

3. The smaller unit of measure may be written as a _____ part of the larger unit of measure.

Ground chuck steak is on sale at the supermarket. The unit price has accidentally been torn off the package, and only the second half of the label remains.

ce	Retail Price
592	$3.11
er lb.	1.2 lb.

Using the information still visible on the meat label, find the unit price of a pound of ground chuck steak.

Solution: (1) Use this relationship:

$$\text{Unit price} = \frac{\text{retail price}}{\text{measure}}.$$

$$\text{Unit price} = \frac{\$3.11}{1.2}.$$

```
                              2.5916   quotient
            divisor   1.2,)3.1,1000   dividend
                              2 4 xxxx
                              7 1
                              6 0
                              1 10
                              1 08
                                20
                                12
                                80
                                72
```

(2) Move the decimal point in the divisor 1.2 *one* place to the right. Move the decimal point in the dividend *one* place to the right. After moving the decimal point in the dividend, place the decimal point directly above it in the quotient.

(3) Rounding off $2.5916 to three decimal places results in $2.592 per lb. Thus the unit price is $2.592 to three decimal places, and the retail price is $3.11 for 1.2 lb.

1. Divide, being sure to move the decimal point the necessary number of places. Round off each answer to three decimal places.

a. $1.02 ÷ 0.34

b. $2.36 ÷ 4.12

c. $1.83 ÷ 0.87

d. $4.79 ÷ 0.12

e. $3.15 ÷ 1.5

f. $0.63 ÷ 1.2

g. $6.43 ÷ 2.32

h. $4.33 ÷ 1.75

i. $14.90 ÷ 5.6

j. $10.06 ÷ 4.38

2. Find the unit price of each item. Round off the unit price to three decimal places.

a.

Unit Price	Retail Price
?	$2.38
per lb.	2 lb.

f.

Unit Price	Retail Price
?	$0.79
per qt.	8 oz.

b.

Unit Price	Retail Price
?	$1.97
per lb.	1.3 lb.

g.

Unit Price	Retail Price
?	$0.95
per lb.	½ lb.

c.

Unit Price	Retail Price
?	$0.89
per lb.	4 oz.

h.

Unit Price	Retail Price
?	$1.12
per lb.	0.83 lb.

d.

Unit Price	Retail Price
?	$3.27
per lb.	1.67 lb.

i.

Unit Price	Retail Price
?	$1.68
per gal.	¼ gal.

e.

Unit Price	Retail Price
?	$0.49
per gal.	per pt.

j.

Unit Price	Retail Price
?	$3.35
per lb.	2 lb. 4 oz.

3. Mrs. Taylor picked up a turkey at the butcher. The turkey cost $20.28 and was priced at $1.69 a pound. How much did the turkey weigh?

4. A bag of jelly beans costs $1.33. If jelly beans cost $1.69 a pound, what is the weight of the bag?

5. A box of nails costs $0.80. Nails are sold for $1.60 a pound. What is the weight of the box? What should the change be from a $5 bill?

EXTRA FOR EXPERTS

1. Mrs. Taylor's turkey (Exercise 3 above) needs 20 minutes a pound to cook. What will be the total cooking time? If Mrs. Taylor stuffs the turkey, it will take 25 minutes a pound to cook. How much more time will the turkey take to cook if stuffed?

2. Mrs. Taylor bought twice as many pounds of potatoes as of onions. She paid $0.87 for onions, which were $0.29 a pound. What was the weight of the onions that she bought? How many pounds of potatoes did she buy? At $0.43 a pound, what was the cost of the potatoes? What was the total cost of the vegetables?

5-5 *Computing the Measure*

STUDY GUIDE   You can find the *measure* of an item if the unit price and the retail price are known. The measure can be found by dividing the retail price by the unit price. The measure of each must be the same before dividing.

$$\text{Measure} = \frac{\text{retail price}}{\text{unit price}}$$

or

$$M = \frac{RP}{UP}.$$

FACT FINDING

1. To find the measure of an item you divide the _____ price by the unit price.

2. The measure of both prices must be the same before _____ .

MODEL PROBLEM

How many pounds does a chicken weigh if the unit price is $0.59 per lb. and the retail price is $3.14?

Solution: (1)   Express the weight as follows:

$$\text{Weight} = \frac{\text{retail price}}{\text{unit price}}.$$

$$\text{Weight} = \frac{\$3.14}{\$0.59}.$$

(2) Move the decimal point in the divisor, $0.59, *two* places to the right. Move the decimal point in the dividend, $3.14, *two* places to the right. Place the decimal point in the quotient directly above the decimal point in the dividend. Divide.

```
            5.322
$0.59. )3.14,000
         2 95
         19 0
         17 7
          1 30
          1 18
            120
            118
```

(3) Rounding off 5.322 to the nearest hundredth pound results in 5.32 lb.

Find each measure to the nearest hundredth.

a.

Unit Price	Retail Price
$1.90 per lb.	$0.95 ? lb.

f.

Unit Price	Retail Price
$1.09 per lb.	$0.87 ? lb.

b.

Unit Price	Retail Price
$0.49 per lb.	$0.98 ? lb.

g.

Unit Price	Retail Price
$0.39 per lb.	$1.53 ? lb.

c.

Unit Price	Retail Price
$5.19 per gal.	$1.73 ? gal.

h.

Unit Price	Retail Price
$0.98 per 4 oz.	$1.96 ? oz.

d.

Unit Price	Retail Price
$0.78 per 2 oz.	$0.39 ? oz.

i.

Unit Price	Retail Price
$0.67 per lb.	$1.39 ? lb.

e.

Unit Price	Retail Price
$2.29 per lb.	$1.43 ? lb.

j.

Unit Price	Retail Price
$2.29 per lb.	$0.63 ? lb.

5-6 Relationships among Unit Price, Measure, and Retail Price

STUDY GUIDE To calculate the unit price, measure, or retail price of an item, you must know the other two values. The following relationships are used.

Retail price = unit price × measure

$$\text{Unit price} = \frac{\text{retail price}}{\text{measure}}$$

$$\text{Measure} = \frac{\text{retail price}}{\text{unit price}}$$

1. Complete the chart. Find unit price to three decimal places, measure to the nearest hundredth, and retail price to the nearest cent.

	Item	Unit Price	Measure	Retail Price
a.	Pancake mix	$0.495/lb.	2 lb.	?
b.	Peanut butter	$1.39/lb.	8 oz.	?
c.	Baked beans	$0.175/8 oz.	?	35¢
d.	Coffee	?/oz.	12 oz.	$3.19
e.	Cocoa mix	?/lb.	12 oz.	$1.29
f.	Cookies	?/oz.	6 oz.	$1.18
g.	Flour	$1.11/5 lb.	2 lb.	?
h.	Apples	?/lb.	1.3 lb.	$1.29
i.	Tea bags	$1.09/100	125	?
j.	Hot dogs	$2.39/lb.	?	$3.59
k.	Chuck steak	$2.19/lb.	1.54 lb.	?
l.	Hot sausage	?/lb.	12 oz.	$1.69
m.	Chicken	$0.79/lb.	?	$3.23
n.	Sirloin steak	?/lb.	2.15 lb.	$7.07

2. Which is the "better" buy?

a. Ice cream A costing $1.79 for a quart or ice cream B costing $3.29 for $\frac{1}{2}$ gal.

b. Corn oil C costing 99¢ for a pint or corn oil D costing $1.49 for 24 oz.

EXTRA FOR EXPERTS

In the rule: Retail price = unit price × measure, if the unit price is doubled and the measure is multiplied by 3 (tripled), what multiple is the resulting retail price of the original?

5-7 *Computing Wages*

STUDY GUIDE A worker employed by a company is called an *employee* of that company. Employees may be full-time or part-time workers.

Full-time employees usually are paid a larger hourly wage. They also receive additional benefits: sick days, vacation time, lunch breaks, and so on, which are not part of their specified salary.

Students are frequently part-time employees since they attend school and work after school hours. Part-time workers usually earn money only for the hours they work.

To compute the *salary* or *wages* of an employee, you multiply the number of hours he or she works by the hourly wage. The product is called the *gross wages*.

Gross wages = number of hours worked × hourly wage

Several taxes are deducted from an employee's wages: Social Security (F.I.C.A.), federal, state, and city taxes. Not all states or cities levy a tax,

and the tax rate varies from one state or city to the next. After these taxes are subtracted from the gross pay, what is left is called *take-home* or *net pay*.

The amount of the tax deductions depends on the rate and the number of exemptions claimed. A person who declares no exemptions pays a higher rate of taxes than a person with one or more exemptions. A married couple with five children has seven deductions. The husband will declare seven deductions and pay a much lower rate of federal, state, and city taxes than the student with no or one exemption.

Tax charts include the number of exemptions, the pay period, the wages, and the tax to be withheld.

FACT FINDING

1. An _____ is someone who works for a company or business.

2. An hourly wage is salary earned per _____ .

3. To compute the gross salary, you _____ the number of hours

worked by the _____ wage.

4. Gross wages minus taxes equals _____ pay.

5. Tax deductions from wages _____ from state to state.

MODEL PROBLEM

Chuck Witsell is a part-time stock boy in the S and T Supermarket. If he earns $3.45 an hour, and works 3, 3, $4\frac{1}{2}$, 3, 3, and 8 hours on Monday through Saturday, respectively, what is his gross salary for the week?

Solution: (1)   Find the total number of hours worked:

$$3 + 3 + 4\frac{1}{2} + 3 + 3 + 8 = 24\frac{1}{2}.$$

(2)   Express $\frac{1}{2}$ hr. as a decimal by dividing the denominator (2) into the numerator (1):

$$2\overline{)1.00} \quad 0.5$$

Thus: $\frac{1}{2}$ hr. = 0.5 hr.

Any fraction can be changed to a decimal in this manner.

(3)   Therefore:

Gross wages = number of hours worked × hourly wages.

Gross wages = $24\frac{1}{2}$ × $3.45

= 24.5 × $3.45

= $84.525

= $84.53

$$\begin{array}{r} \$3.45 \\ \times\ 24.5 \\ \hline 1\ 725 \\ 13\ 80 \\ 69\ 0 \\ \hline \$84.525 \end{array}$$

(4)   Chuck is paid on a weekly basis and declares 0 exemption. Chuck's gross pay is $84.53. Compute his deductions as follows:

Refer to Table 1 (N.Y. State Tax) =	$2.10
Refer to Table 2 (N.Y. City Tax) =	0.95
Refer to Table 3 (Federal Tax) =	9.30
Social Security (F.I.C.A.; =	5.38
this is obtained from a table)	
Total deductions =	$17.73

Chuck's net wages = gross wages − deductions.

Gross wages =	$84.53
Deductions =	− 17.73
Net wages =	$66.80

Table 1
N.Y. State—Weekly Payroll Period

WAGES At Least	But Less Than	0	1	2	3	4	5
		TAX TO BE WITHHELD					
$ 0	$16	$.00					
16	17	.10					
17	18	.10					
18	19	.10					
19	20	.10					
20	21	.10					
21	22	.20					
22	23	.20					
23	24	.20					
24	25	.20					
25	26	.20					
26	27	.30					
27	28	.30					
28	29	.30					
29	30	.30					
30	31	.30	$.10				
31	32	.40	.10				
32	33	.40	.10				
33	34	.40	.10				
34	35	.40	.10				
35	36	.50	.20				
36	37	.50	.20				
37	38	.50	.20				
38	39	.60	.20				
39	40	.60	.20				
40	41	.60	.30				
41	42	.60	.30				
42	43	.70	.30				
43	44	.70	.30				
44	45	.70	.30				
45	46	.80	.40	$.10			
46	47	.80	.40	.10			
47	48	.80	.40	.10			
48	49	.90	.40	.10			
49	50	.90	.50	.10			
50	51	.90	.50	.20			
51	52	.90	.50	.20			
52	53	1.00	.50	.20			
53	54	1.00	.60	.20			
54	55	1.00	.60	.20			
55	56	1.10	.60	.30			
56	57	1.10	.70	.30			
57	58	1.10	.70	.30			
58	59	1.20	.70	.30			
59	60	1.20	.80	.30	$.10		
60	62	1.20	.80	.40	.10		
62	64	1.30	.90	.40	.10		
64	66	1.40	.90	.50	.20		
66	68	1.40	1.00	.50	.20		
68	70	1.50	1.00	.60	.20		
70	72	1.50	1.10	.70	.30		
72	74	1.60	1.20	.70	.30		
74	76	1.70	1.20	.80	.40	$.10	
76	78	1.80	1.30	.80	.40	.10	
78	80	1.90	1.30	.90	.50	.20	
80	82	1.90	1.40	1.00	.50	.20	
82	84	2.00	1.50	1.00	.60	.20	
84	86	2.10	1.50	1.10	.70	.30	
86	88	2.20	1.60	1.10	.70	.30	
88	90	2.30	1.70	1.20	.80	.40	$.10
90	92	2.30	1.80	1.30	.80	.40	.10

WAGES At Least	But Less Than	0	1	2	3	4	5	6	7	8	9	10 or more
		TAX TO BE WITHHELD										
$92	$94	$2.40	$1.80	$1.30	$ .90	$ .50	$ .10					
94	96	2.50	1.90	1.40	1.00	.50	.20					
96	98	2.60	2.00	1.40	1.00	.60	.20					
98	100	2.70	2.10	1.50	1.10	.60	.30					
100	105	2.80	2.20	1.60	1.20	.70	.30	$.10				
105	110	3.00	2.40	1.80	1.30	.90	.50	.20				
110	115	3.20	2.60	2.00	1.50	1.00	.60	.30				
115	120	3.50	2.80	2.20	1.70	1.20	.80	.40	$ .10			
120	125	3.70	3.00	2.40	1.90	1.30	.90	.50	.20			
125	130	4.00	3.30	2.60	2.10	1.50	1.10	.60	.30			
130	135	4.20	3.50	2.80	2.30	1.70	1.20	.80	.40	$ .10		
135	140	4.50	3.80	3.00	2.50	1.90	1.40	.90	.50	.20		
140	145	4.70	4.00	3.30	2.70	2.10	1.50	1.10	.70	.30		
145	150	5.00	4.30	3.50	2.90	2.30	1.70	1.20	.80	.40	$ .10	
150	160	5.40	4.60	3.90	3.20	2.60	2.00	1.50	1.00	.60	.20	
160	170	6.00	5.20	4.40	3.70	3.00	2.40	1.80	1.30	.90	.50	$ .10
170	180	6.60	5.70	4.90	4.10	3.40	2.80	2.20	1.60	1.20	.70	.30
180	190	7.10	6.20	5.30	4.50	3.80	3.10	2.50	1.90	1.40	1.00	.50
190	200	7.60	6.70	5.80	5.00	4.20	3.50	2.90	2.30	1.70	1.20	.80
200	210	8.20	7.20	6.30	5.50	4.70	3.90	3.20	2.60	2.00	1.50	1.10
210	220	8.80	7.80	6.80	6.00	5.10	4.40	3.60	3.00	2.40	1.80	1.30
220	230	9.40	8.40	7.30	6.50	5.60	4.80	4.10	3.30	2.70	2.10	1.60
230	240	10.00	8.90	7.90	7.00	6.10	5.30	4.50	3.80	3.10	2.50	1.90
240	250	10.60	9.50	8.50	7.50	6.60	5.80	4.90	4.20	3.50	2.80	2.20
250	260	11.30	10.10	9.10	8.10	7.10	6.30	5.40	4.60	3.90	3.20	2.60
260	270	12.00	10.80	9.70	8.70	7.70	6.80	5.90	5.00	4.30	3.60	2.90
270	280	12.60	11.50	10.30	9.30	8.30	7.30	6.40	5.50	4.70	4.00	3.30
280	290	13.30	12.20	11.00	9.90	8.90	7.80	6.90	6.00	5.20	4.40	3.70
290	300	14.20	12.90	11.80	10.60	9.50	8.50	7.50	6.60	5.70	4.90	4.20
300	310	15.10	13.80	12.60	11.40	10.20	9.20	8.20	7.20	6.30	5.50	4.70
310	320	16.00	14.70	13.40	12.20	11.00	9.90	8.90	7.90	6.90	6.10	5.20
320	330	16.90	15.60	14.30	13.00	11.80	10.70	9.60	8.60	7.60	6.70	5.80
330	340	17.90	16.50	15.20	13.90	12.60	11.50	10.30	9.30	8.30	7.30	6.40
340	350	18.90	17.50	16.10	14.80	13.50	12.30	11.10	10.00	9.00	8.00	7.00
350	360	19.90	18.50	17.00	15.70	14.40	13.10	11.90	10.80	9.70	8.70	7.70
360	370	21.00	19.50	18.00	16.60	15.30	14.00	12.70	11.60	10.40	9.40	8.40
370	380	22.10	20.50	19.00	17.60	16.20	14.90	13.60	12.40	11.20	10.10	9.10
380	390	23.20	21.60	20.00	18.60	17.20	15.80	14.50	13.20	12.00	10.90	9.80
390	400	24.30	22.70	21.10	19.60	18.20	16.70	15.40	14.10	12.80	11.70	10.50
400	410	25.40	23.80	22.20	20.60	19.20	17.70	16.30	15.00	13.70	12.50	11.30
410	420	26.50	24.90	23.30	21.70	20.20	18.70	17.30	15.90	14.60	13.30	12.10
420	430	27.60	26.00	24.40	22.80	21.20	19.70	18.30	16.80	15.50	14.20	12.90
430	440	28.70	27.10	25.50	23.90	22.30	20.70	19.30	17.80	16.40	15.10	13.80
440	450	29.80	28.20	26.60	25.00	23.40	21.80	20.30	18.80	17.40	16.00	14.70
450	460	30.90	29.30	27.70	26.10	24.50	22.90	21.40	19.80	18.40	16.90	15.60
460	470	32.00	30.40	28.80	27.20	25.60	24.00	22.50	20.90	19.40	17.90	16.50
470	480	33.10	31.50	29.90	28.30	26.70	25.10	23.60	22.00	20.40	18.90	17.50
480	490	34.20	32.60	31.00	29.40	27.80	26.20	24.70	23.10	21.50	19.90	18.50
490	500	35.30	33.70	32.10	30.50	28.90	27.30	25.80	24.20	22.60	21.00	19.50
500	510	36.40	34.80	33.20	31.60	30.00	28.40	26.90	25.30	23.70	22.10	20.50
510	520	37.50	35.90	34.30	32.70	31.10	29.50	28.00	26.40	24.80	23.20	21.60
520	530	38.60	37.00	35.40	33.80	32.20	30.60	29.10	27.50	25.90	24.30	22.70
530	540	39.70	38.10	36.50	34.90	33.30	31.70	30.20	28.60	27.00	25.40	23.80
540	550	40.80	39.20	37.60	36.00	34.40	32.80	31.30	29.70	28.10	26.50	24.90
550	560	41.90	40.30	38.70	37.10	35.50	33.90	32.40	30.80	29.20	27.60	26.00
560	570	43.00	41.40	39.80	38.20	36.60	35.00	33.50	31.90	30.30	28.70	27.10
570	580	44.10	42.50	40.90	39.30	37.70	36.10	34.60	33.00	31.40	29.80	28.20
580	590	45.20	43.60	42.00	40.40	38.80	37.20	35.70	34.10	32.50	30.90	29.30
590	600	46.30	44.70	43.10	41.50	39.90	38.30	36.80	35.20	33.60	32.00	30.40
$600 & over		11 percent of the excess over $600 plus —										
		46.80	45.20	43.70	42.10	40.50	38.90	37.30	35.70	34.10	32.50	31.00

Table 2
N.Y. City—Resident Tax
Weekly Payroll Period

WAGES At Least	But Less Than	0	1	2	3	4	5
		TAX TO BE WITHHELD					
$0	$16	$.00					
16	17	.05					
17	18	.05					
18	19	.05					
19	20	.05					
20	21	.05					
21	22	.05					
22	23	.10					
23	24	.10					
24	25	.10					
25	26	.10					
26	27	.10					
27	28	.15					
28	29	.15					
29	30	.15					
30	31	.15					
31	32	.15	$.05				
32	33	.15	.05				
33	34	.20	.05				
34	35	.20	.05				
35	36	.20	.05				
36	37	.25	.10				
37	38	.25	.10				
38	39	.25	.10				
39	40	.25	.10				
40	41	.30	.10				
41	42	.30	.10				
42	43	.30	.15				
43	44	.30	.15				
44	45	.35	.15				
45	46	.35	.15	$.05			
46	47	.35	.15	.05			
47	48	.40	.20	.05			
48	49	.40	.20	.05			
49	50	.40	.20	.05			
50	51	.40	.20	.05			
51	52	.45	.25	.10			
52	53	.45	.25	.10			
53	54	.45	.25	.10			
54	55	.50	.30	.10			
55	56	.50	.30	.10			
56	57	.50	.30	.15			
57	58	.50	.30	.15			
58	59	.55	.35	.15			
59	60	.55	.35	.15	$.05		
60	62	.55	.35	.15	.05		
62	64	.60	.40	.20	.05		
64	66	.65	.40	.20	.05		
66	68	.65	.45	.25	.10		
68	70	.70	.50	.30	.10		
70	72	.70	.50	.30	.15		
72	74	.75	.55	.35	.15		
74	76	.80	.55	.35	.15	$.05	
76	78	.80	.60	.40	.20	.05	
78	80	.85	.60	.40	.20	.05	
80	82	.90	.65	.45	.25	.10	
82	84	.90	.70	.45	.25	.10	
84	86	.95	.70	.50	.30	.10	
86	88	1.00	.75	.55	.35	.15	
88	90	1.05	.75	.55	.35	.15	$.05
90	92	1.05	.80	.60	.40	.20	.05

WAGES At Least	But Less Than	0	1	2	3	4	5	6	7	8	9	10 or more
		TAX TO BE WITHHELD										
$92	$94	$1.10	$.85	$.60	$.40	$.20	$.05					
94	96	1.15	.90	.65	.45	.25	.10					
96	98	1.20	.90	.65	.45	.25	.10					
98	100	1.20	.95	.70	.50	.30	.10					
100	105	1.30	1.00	.75	.55	.35	.15					
105	110	1.35	1.10	.85	.60	.40	.20	$.05				
110	115	1.45	1.20	.95	.70	.50	.30	.10				
115	120	1.55	1.30	1.05	.75	.55	.35	.15	$.05			
120	125	1.65	1.40	1.10	.85	.60	.40	.20	.05			
125	130	1.75	1.45	1.20	.95	.70	.50	.30	.10			
130	135	1.85	1.55	1.30	1.05	.80	.55	.35	.15	$.05		
135	140	1.95	1.65	1.40	1.15	.85	.65	.45	.25	.10		
140	145	2.05	1.75	1.50	1.20	.95	.70	.50	.30	.10		
145	150	2.15	1.85	1.60	1.30	1.05	.80	.55	.35	.15	$.05	
150	160	2.35	2.00	1.75	1.45	1.20	.90	.65	.45	.25	.10	
160	170	2.55	2.25	1.95	1.65	1.35	1.10	.85	.60	.40	.20	$.05
170	180	2.75	2.45	2.10	1.80	1.55	1.25	1.00	.75	.55	.35	.15
180	190	2.95	2.65	2.30	2.00	1.70	1.40	1.15	.90	.65	.45	.25
190	200	3.15	2.80	2.50	2.15	1.85	1.60	1.30	1.05	.80	.55	.35
200	210	3.35	3.00	2.70	2.35	2.05	1.75	1.45	1.20	.95	.70	.50
210	220	3.60	3.20	2.90	2.55	2.20	1.90	1.65	1.35	1.10	.85	.60
220	230	3.80	3.45	3.05	2.75	2.40	2.10	1.80	1.50	1.25	1.00	.70
230	240	4.00	3.65	3.30	2.95	2.60	2.25	1.95	1.70	1.40	1.15	.85
240	250	4.25	3.85	3.50	3.15	2.80	2.45	2.15	1.85	1.55	1.30	1.00
250	260	4.45	4.05	3.70	3.35	3.00	2.65	2.30	2.00	1.75	1.45	1.20
260	270	4.70	4.30	3.90	3.55	3.20	2.85	2.50	2.20	1.90	1.60	1.35
270	280	4.90	4.50	4.15	3.75	3.40	3.05	2.70	2.40	2.05	1.75	1.50
280	290	5.15	4.75	4.35	3.95	3.60	3.25	2.90	2.55	2.25	1.95	1.65
290	300	5.40	5.00	4.60	4.20	3.85	3.50	3.15	2.80	2.45	2.15	1.85
300	310	5.70	5.30	4.90	4.50	4.10	3.75	3.40	3.00	2.70	2.35	2.05
310	320	6.00	5.60	5.15	4.75	4.35	4.00	3.65	3.25	2.90	2.60	2.25
320	330	6.30	5.85	5.45	5.05	4.65	4.25	3.90	3.50	3.15	2.80	2.50
330	340	6.60	6.15	5.75	5.30	4.90	4.50	4.15	3.75	3.40	3.05	2.70
340	350	6.90	6.45	6.05	5.60	5.20	4.80	4.40	4.00	3.65	3.30	2.95
350	360	7.20	6.80	6.35	5.90	5.50	5.05	4.65	4.30	3.90	3.55	3.20
360	370	7.55	7.10	6.65	6.20	5.75	5.35	4.95	4.55	4.15	3.80	3.45
370	380	7.85	7.40	6.95	6.50	6.05	5.65	5.25	4.80	4.45	4.05	3.70
380	390	8.20	7.75	7.25	6.80	6.35	5.95	5.50	5.10	4.70	4.30	3.95
390	400	8.55	8.05	7.60	7.10	6.65	6.25	5.80	5.40	5.00	4.60	4.20
400	410	8.90	8.40	7.90	7.45	7.00	6.55	6.10	5.70	5.25	4.85	4.45
410	420	9.25	8.70	8.25	7.75	7.30	6.85	6.40	5.95	5.55	5.15	4.75
420	430	9.60	9.05	8.55	8.10	7.60	7.15	6.70	6.25	5.85	5.40	5.00
430	440	9.95	9.40	8.90	8.45	7.95	7.45	7.00	6.55	6.15	5.70	5.30
440	450	10.30	9.75	9.25	8.75	8.30	7.80	7.35	6.90	6.45	6.00	5.60
450	460	10.70	10.15	9.60	9.10	8.60	8.15	7.65	7.20	6.75	6.30	5.85
460	470	11.05	10.50	9.95	9.45	8.95	8.45	8.00	7.50	7.05	6.60	6.15
470	480	11.45	10.90	10.35	9.80	9.30	8.80	8.30	7.85	7.35	6.90	6.45
480	490	11.85	11.25	10.75	10.20	9.65	9.15	8.65	8.15	7.70	7.25	6.80
490	500	12.25	11.65	11.10	10.55	10.00	9.50	9.00	8.50	8.05	7.55	7.10
500	510	12.65	12.05	11.50	10.95	10.40	9.85	9.35	8.85	8.35	7.90	7.40
510	520	13.05	12.45	11.90	11.30	10.75	10.20	9.70	9.20	8.70	8.20	7.75
520	530	13.50	12.85	12.30	11.70	11.15	10.60	10.05	9.55	9.05	8.55	8.05
530	540	13.90	13.30	12.70	12.10	11.55	11.00	10.45	9.90	9.40	8.90	8.40
540	550	14.35	13.70	13.10	12.50	11.95	11.35	10.80	10.25	9.75	9.25	8.75
550	560	14.75	14.15	13.55	12.90	12.35	11.75	11.20	10.65	10.10	9.60	9.10
560	570	15.20	14.60	13.95	13.35	12.75	12.15	11.60	11.05	10.50	9.95	9.45
570	580	15.65	15.00	14.40	13.75	13.15	12.55	12.00	11.40	10.85	10.30	9.80
580	590	16.05	15.45	14.80	14.20	13.60	12.95	12.40	11.80	11.25	10.70	10.15
590	600	16.50	15.85	15.25	14.65	14.00	13.40	12.80	12.20	11.65	11.05	10.50
$600 & over		4.3 percent of the excess over $600 plus —										
		16.70	16.10	15.45	14.85	14.20	13.60	13.00	12.40	11.85	11.25	10.70

Table 3
Federal Tax
Single Persons—Weekly Payroll Period
(The wavy lines indicate that
part of the table has been omitted.)

And the wages are—		And the number of withholding allowances claimed is—										
At least	But less than	0	1	2	3	4	5	6	7	8	9	10 or more
		The amount of income tax to be withheld shall be—										
$0	$28	$0	$0	$0	$0	$0	$0	$0	$0	$0	$0	$0
28	29	.20	0	0	0	0	0	0	0	0	0	0
29	30	.30	0	0	0	0	0	0	0	0	0	0
30	31	.50	0	0	0	0	0	0	0	0	0	0
31	32	.60	0	0	0	0	0	0	0	0	0	0
57	58	4.50	1.60	0	0	0	0	0	0	0	0	0
58	59	4.70	1.80	0	0	0	0	0	0	0	0	0
59	60	4.80	1.90	0	0	0	0	0	0	0	0	0
60	62	5.10	2.20	0	0	0	0	0	0	0	0	0
62	64	5.40	2.50	0	0	0	0	0	0	0	0	0
64	66	5.70	2.80	0	0	0	0	0	0	0	0	0
66	68	6.10	3.10	.20	0	0	0	0	0	0	0	0
68	70	6.40	3.40	.50	0	0	0	0	0	0	0	0
70	72	6.80	3.70	.80	0	0	0	0	0	0	0	0
72	74	7.10	4.00	1.10	0	0	0	0	0	0	0	0
74	76	7.50	4.30	1.40	0	0	0	0	0	0	0	0
76	78	7.90	4.60	1.70	0	0	0	0	0	0	0	0
78	80	8.20	4.90	2.00	0	0	0	0	0	0	0	0
80	82	8.60	5.20	2.30	0	0	0	0	0	0	0	0
82	84	8.90	5.50	2.60	0	0	0	0	0	0	0	0
84	86	9.30	5.80	2.90	0	0	0	0	0	0	0	0
86	88	9.70	6.20	3.20	.30	0	0	0	0	0	0	0
88	90	10.00	6.60	3.50	.60	0	0	0	0	0	0	0
90	92	10.40	6.90	3.80	.90	0	0	0	0	0	0	0
92	94	10.70	7.30	4.10	1.20	0	0	0	0	0	0	0
94	96	11.10	7.60	4.40	1.50	0	0	0	0	0	0	0
96	98	11.50	8.00	4.70	1.80	0	0	0	0	0	0	0
98	100	11.80	8.40	5.00	2.10	0	0	0	0	0	0	0
100	105	12.50	9.00	5.50	2.60	0	0	0	0	0	0	0
105	110	13.40	9.90	6.40	3.40	.50	0	0	0	0	0	0
110	115	14.30	10.80	7.30	4.10	1.20	0	0	0	0	0	0
115	120	15.20	11.70	8.20	4.90	2.00	0	0	0	0	0	0
120	125	16.10	12.60	9.10	5.70	2.70	0	0	0	0	0	0
125	130	17.00	13.50	10.00	6.60	3.50	.60	0	0	0	0	0
130	135	17.90	14.40	10.90	7.50	4.20	1.40	0	0	0	0	0
$135	$140	$19.00	$15.30	$11.80	$8.40	$5.00	$2.10	$0	$0	$0	$0	$0
140	145	20.00	16.20	12.70	9.30	5.80	2.90	0	0	0	0	0
145	150	21.10	17.10	13.60	10.20	6.70	3.60	.70	0	0	0	0
150	160	22.60	18.60	15.00	11.50	8.10	4.70	1.80	0	0	0	0
160	170	24.70	20.70	16.80	13.30	9.90	6.40	3.30	.50	0	0	0
170	180	26.80	22.80	18.80	15.10	11.70	8.20	4.80	2.00	0	0	0
180	190	28.90	24.90	20.90	16.90	13.50	10.00	6.50	3.50	.60	0	0
190	200	31.00	27.00	23.00	18.90	15.30	11.80	8.30	5.00	2.10	0	0
200	210	33.60	29.10	25.10	21.00	17.10	13.60	10.10	6.70	3.60	.70	0
210	220	36.20	31.20	27.20	23.10	19.10	15.40	11.90	8.50	5.10	2.20	0
220	230	38.80	33.80	29.30	25.20	21.20	17.20	13.70	10.30	6.80	3.70	.80
230	240	41.40	36.40	31.40	27.30	23.30	19.20	15.50	12.10	8.60	5.20	2.30
240	250	44.00	39.00	34.00	29.40	25.40	21.30	17.30	13.90	10.40	6.90	3.80
250	260	46.60	41.60	36.60	31.60	27.50	23.40	19.40	15.70	12.20	8.70	5.30
260	270	49.20	44.20	39.20	34.20	29.60	25.50	21.50	17.50	14.00	10.50	7.10
270	280	51.80	46.80	41.80	36.80	31.80	27.60	23.60	19.60	15.80	12.30	8.90
280	290	54.80	49.40	44.40	39.40	34.40	29.70	25.70	21.70	17.60	14.10	10.70
290	300	57.80	52.10	47.00	42.00	37.00	32.00	27.80	23.80	19.70	15.90	12.50
300	310	60.80	55.10	49.60	44.60	39.60	34.60	29.90	25.90	21.80	17.80	14.30
310	320	63.80	58.10	52.30	47.20	42.20	37.20	32.20	28.00	23.90	19.90	16.10
320	330	66.80	61.10	55.30	49.80	44.80	39.80	34.80	30.10	26.00	22.00	17.90
330	340	70.00	64.10	58.30	52.50	47.40	42.40	37.40	32.40	28.10	24.10	20.00
340	350	73.40	67.10	61.30	55.50	50.00	45.00	40.00	35.00	30.20	26.20	22.10
350	360	76.80	70.30	64.30	58.50	52.80	47.60	42.60	37.60	32.60	28.30	24.20
360	370	80.20	73.70	67.30	61.50	55.80	50.20	45.20	40.20	35.20	30.40	26.30

1. Change each fraction to a two-place decimal.

   a. $\frac{1}{4}$ c. $\frac{3}{4}$ e. $\frac{1}{10}$ g. $\frac{1}{3}$ i. $\frac{2}{3}$

   b. $\frac{2}{5}$ d. $\frac{1}{2}$ f. $\frac{3}{5}$ h. $\frac{7}{8}$ j. $\frac{5}{6}$

2. Complete the payroll chart for the part-time employees at the S and T Supermarket.

S and T Supermarket Payroll

	Name	M	T	W	Th	F	S	Sun	Total Hours	Rate per Hour	Gross Wages
a.	Deborah Smith	3	4	$3\frac{1}{2}$	$2\frac{1}{4}$	4	6	—	$22\frac{3}{4}$ = 22.75	$3.45	$78.49
b.	Cathy Southerland	4	$3\frac{1}{2}$	$2\frac{3}{4}$	3	—	—	7		$3.60	
c.	Neddy Homere	$3\frac{1}{2}$	$2\frac{3}{4}$	3	4	3	8	—		$4.10	
d.	Samuel Cox	$2\frac{3}{4}$	3	4	5	5	4	—		$3.45	
e.	Brenda Estrada	3	4	5	3	4	5	—		$3.70	
f.	Aaron Coleman	4	5	3	4	—	—	6		$3.55	
g.	Lydia Bermudez	5	3	4	$3\frac{1}{2}$	$2\frac{1}{2}$	4	—		$3.90	

3. Referring to Table 1, find the amount of state taxes deducted from the following gross wages if the employee has 1 exemption:

   a. $56.25   d. $218.94   g. $176.50

   b. $490.00   e. $376.52   h. $400.00

   c. $274.37   f. $112.37   i. $89.00

4. Referring to Table 2, find the amount of city taxes deducted from the following gross wages if the employee has 3 exemptions:

   a. $73.76   d. $438.75   g. $315.00

   b. $361.25   e. $116.89   h. $142.00

   c. $200.00   f. $89.50   i. $401.35

5. Referring to Table 3, find the amount of federal taxes deducted from the following gross wages if the employee has 2 exemptions:

a. $93.50		**d.** $144.00		**g.** $90.00
b. $65.20		**e.** $71.80		**h.** $84.70
c. $137.50		**f.** $159.00		**i.** $166.00

6. If the taxes to be deducted are as indicated, find the net wage in each example.

	Gross Wages	Taxes	Net Wages
a.	$56.25	$5.73	
b.	$490.00	$175.28	
c.	$274.00	$69.75	
d.	$218.94	$47.25	
e.	$376.52	$153.18	
f.	$89.00	$14.75	

7. If the taxes deducted are as indicated and the net wages as computed, find the gross wages and the hourly rate if a 40-hour week was completed.

	Hourly Rate	Gross Wages	Taxes	Net Wages
a.			$29.00	$165.35
b.			$47.65	$208.10
c.			$162.00	$326.00
d.			$8.65	$75.50
e.			$27.46	$110.75
f.			$72.18	$165.34
g.			$201.40	$436.00

8. Sammy works after school every day for 3 hours at Burger Mart. On Saturday he works $7\frac{1}{2}$ hours. What are his gross wages for the week if he earns $3.40 an hour?

9. For the summer Sammy was told that he could work a full week including Saturday from 8:00 A.M. till 4:00 P.M. (He has a half-hour lunch break that he does not get paid for.) How many hours a week will he get paid for? At $3.45 an hour, what will his gross wages be?

10. If Francesca has a gross pay of $210 per week, what are her yearly gross wages?

11. If Johnson has gross wages of $16,000, what is his weekly gross pay?

12. Jackson gets paid monthly and his gross monthly wages are $650. What are his weekly gross wages? What are his annual gross wages?

13. Arlene gets paid bimonthly, and her net pay is $275.50. What is her yearly net pay?

14. Thomas worked at the supermarket and earned $17.50 per day. He quit and obtained a job in the local movie house at $22.00 per day. How much more is he earning per week?

5-8 *Punching a Time Card*

STUDY GUIDE Payrolls are usually based on the number of hours worked. This information is recorded on the employee's time card, which is punched daily. Arrivals and departures to and from work and for lunch are recorded.

With this information, the bookkeeper is able to determine the number of hours worked, which, along with the hourly rate of pay, is the basis for the employee's gross wages. Time cards are kept as official records and may be audited or checked by the boss or an outside agency.

FACT FINDING

1. The payroll is completed by the _____ .

2. Each employee _____ a time card daily.

3. The time card records the time of arrival to work and _____ from work.

4. The _____ wages of an employee depend on the hourly rate of pay and the number of hours worked.

5. An employer may verify an employee's arrival time by checking his or her _____ _____ .

MODEL PROBLEM

Annie arrived for work at 8:00 A.M., left for lunch at 11:30 A.M., returned from lunch at 12:30 P.M., and left for home at 4:30 P.M. What are Annie's gross daily wages if her hourly wage is $3.45?

Solution: (1) Subtract Annie's arrival time from the start of her lunch break. All A.M. times are based on times after midnight.

Thus: 11:30 A.M. means 11 hr. 30 min. after midnight
       − 8:00 A.M. means   8 hr.   0 min. after midnight
                            3 hr. 30 min.

Since 60 minutes = 1 hour, $\frac{30}{60} = \frac{1}{2}$ hr. Thus Annie

worked for 3 hr. 30 min. or $3\frac{1}{2}$ hr. before leaving for lunch.

(2) Subtract Annie's returning time after lunch from her departure time for the day. All P.M. times are based on time after noon.

Thus: 4:30 P.M. means 4 hr. 30 min. after noon
        − 12:30 P.M. means 0 hr. 30 min. after noon
                           4 hr.   0 min. is the time
                                          worked after
                                          lunch.

(3) Therefore the total number of hours worked all day is

$$3\frac{1}{2}$$
$$+ 4$$
$$7\frac{1}{2} \text{ hr.}$$

(4) Gross wages = number of hours worked × hourly wage.

$= 7\frac{1}{2} \times \$3.45$ $\$3.45$
$= 7.5 \times \$3.45$ $\underline{\times\ 7.5}$
$= \$25.875$ $1\ 725$
$= \$25.88$ $\underline{24\ 15}$
                            $\$25.875$

EXERCISES

1. Express each time as hours and minutes after midnight or after noon.

   a. 10:15 A.M.   Answer: 10 hr. 15 min. after midnight

   b. 2:30 P.M.   Answer: 2 hr. 30 min. after noon

   c. 8:15 P.M.   **g.** 5:30 P.M.

   d. 11:45 A.M.   **h.** 9: 30 A.M.

   e. 4:00 A.M.   **i.** 7:45 P.M.

   f. 12:15 P.M.   **j.** 6:18 A.M.

2. Rename each of the times below. Follow the sample shown.

   a. 5 hr. 15 min. = 4 hr. __ min.
      Answer: Since 1 hr. = 60 min.,
              5 hr. 15 min. = 4 hr. + 1 hr. + 15 min.
                            = 4 hr. + 60 min. + 15 min.
                            = 4 hr. 75 min.

   b. 6 hr. 30 min. = 5 hr. __ min.

   c. 8 hr. 10 min. = 7 hr. __ min.

   d. 2 hr. 3 min. = 1 hr. __ min.

   e. 11 hr. 25 min. = 10 hr. __ min.

   f. 4 hr. 20 min. = 3 hr. __ min.

   g. 1 hr. 50 min. = 0 hr. __ min.

   h. 10 hr. 18 min. = 9 hr. __ min.

   i. 9 hr. 45 min. = 8 hr. __ min.

   j. 12 hr. 0 min. = 11 hr. __ min.

3. Subtract:

   a.   4 hr. 30 min.
        − 1 hr. 20 min.

   b.   5 hr. 10 min.
        − 2 hr. 5 min.

   c.   6 hr. 15 min.   =   5 hr. 75 min.
        − 2 hr. 30 min.   − 2 hr. 30 min.
                               3 hr. 45 min.

   d.   5 hr. 30 min.
        − 0 hr. 45 min.

   e.   2 hr. 40 min.
        − 1 hr. 50 min.

   f.   3 hr. 5 min.
        − 0 hr. 40 min.

   g.   11 hr. 15 min.
        − 7 hr. 30 min.

   h.   12 hr.
        − 8 hr. 30 min.

4. Complete the table.

	Name	Arrival	Lunch To	Lunch From	Departure	Hours Worked
a.	Lou Medina	7:30 A.M.	11:30	12:30	5:00 P.M.	
b.	Anna Sanchez	7:30 A.M.	12:00	12:30	4:00 P.M.	
c.	Helen Schmidt	8:00 A.M.	11:00	12:00	5:30 P.M.	
d.	Shirley Nelson	8:00 A.M.	11:30	12:15	5:15 P.M.	
e.	Luis Guzman	7:30 A.M.	11:15	12:15	5:15 P.M.	
f.	Yvette Mercado	9:00 A.M.	1:00	1:30	5:30 P.M.	
g.	John Rivas	8:30 A.M.	12:15	1:00	4:45 P.M.	
h.	Delores Branch	7:00 A.M.	12:00	12:30	4:00 P.M.	
i.	Ana Ramos	7:30 A.M.	11:15	12:30	5:30 P.M.	
j.	Chris Hunter	8:00 A.M.	12:00	1:00	5:15 P.M.	

5. Figure the gross wages per day of each of the employees in the table of Exercise 4, using the hourly rate listed below.

	Employee	Hours Worked per Day	Hourly Rate of Pay	Gross Wages per Day
a.	Lou Medina		$4.10	
b.	Anna Sanchez		3.84	
c.	Helen Schmidt		4.00	
d.	Shirley Nelson		3.50	
e.	Luis Guzman		3.50	
f.	Yvette Mercado		3.65	
g.	John Rivas		4.25	
h.	Delores Branch		5.10	
i.	Ana Ramos		4.25	
j.	Chris Hunter		3.40	

6. Find the weekly gross wages for a 5-day week of each of the employees in the table of Exercise 5.

7. Diane works from 9:00 A.M. till 12:30 and from 1:30 till 5:30 P.M. in a dress shop. She earns $3.90 an hour.

    a. What is her daily gross pay?

    b. What is her weekly gross pay?

    c. If she earns a $2.00 bonus for each dress she sells, what will be her gross pay at the end of a week when she sells 10 dresses?

8. José works from 8:30 A.M. to 12:00 noon and from 1:00 till 4:00 P.M. He earns $4.50 an hour. He is considering a new job at $4.25 an hour with the same working hours and a paid lunch break. Which job offers the better gross wages?

9. In order to be able to save money Mr. Brown works two jobs. He works in an office from 8:30 till 12:30 and from 1:00 till 4:30 at $5.65 an hour. After dinner he works in a garage from 6:00 till midnight at $4.80 an hour.

    a. What are his daily gross wages from the office job?

    b. What are his weekly gross wages?

    c. What are his daily gross wages from the job at the garage?

    d. What are his weekly gross wages for a 5-day week?

    e. If a total of $52.95 is deducted from his weekly office check and $36 is deducted from his weekly garage job, how much is he bringing home each week?

10. Mrs. Holmes pays a babysitter to watch her daughter while she works. She works from 8:00 A.M. till 12:00 noon and from 1:00 P.M. till 4:00 P.M. at $6.25 an hour. She pays the babysitter from 7:30 A.M. till 4:30 P.M. at $2.25 an hour. Deducting what Mrs. Holmes pays the babysitter, calculate her daily gross wages.

1. Jones works for a meat packer. He earns $6.10 per hour. He works 8 to 4 daily, 7 to noon on Saturday, and gets paid for lunch. He was $\frac{1}{2}$ hour late Monday and was docked at his usual pay rate. What will he gross this week?

2. Francine earns $5.50 per hour and works a 40-hour week. She is not paid if she is absent. She is ill Monday and plans to work extra hours Tuesday through Friday to compensate for her absence. How many hours must she work each day to make up for her absence?

3. The owner of a beauty salon finds that she must gross an amount double her employee's salary to cover overhead expenses. If Sally works a 50-hour week at $5.00 per hour, what must be the business receipts to cover expenses? Do you think that overhead should equal the wages of an employee? How may the owner increase profits?

5-9 *Salary (Hourly and Overtime Wages)*

**STUDY
GUIDE**
Many employees' salaries are based on an *hourly rate* of pay. If an employee's usual work week is 35 hours, *overtime* is earned for all hours worked above 35 hours.

The number of hours in a regular work week is usually set by the employer or by union contract. The standard overtime rate of pay is one-and-a-half times the regular hourly pay. This is called *time-and-a-half*. On some jobs, overtime on Saturdays and Sundays and after midnight is two times the regular hourly pay. This is called *double time*.

FACT
FINDING

1. Standard overtime rate of pay is usually _____ times the regular hourly rate of pay.

2. _____ hours is a usual number of regular hours for a week for an employee.

3. _____ is one-and-a-half times the regular hourly rate of pay.

4. Overtime on some jobs for Saturdays and Sundays is called _____ _____ .

Ruth earns $5.60 per hour. What is her weekly salary if she works 2 hours of overtime beyond her regular 35-hour work week?

Solution: (1) Computing her regular salary, you multiply hourly rate by regular hours:

$$\begin{array}{r} \$5.60 \\ \times\ 35 \\ \hline 28\ 00 \\ 168\ 0 \\ \hline \$196.00 \end{array}$$

(2) Overtime pay is $1\frac{1}{2}$ times the regular hourly rate. Therefore:

$$1\frac{1}{2} \times \$5.60 \qquad\qquad\qquad or \qquad 1.5 \times \$5.60$$

$$\frac{\overset{3}{\cancel{\frac{3}{2}}}}{\underset{1}{}} \times \frac{\overset{2.80}{\cancel{\$.60}}}{1} = \$8.40 \text{ per hour} \qquad\qquad \begin{array}{r} \$5.60 \\ \times\ 1.5 \\ \hline 2\ 800 \\ 5\ 60 \\ \hline \$8.40\cancel{0} \end{array}$$

2 hours of overtime is
$$\begin{array}{r} \$8.40 \\ \times\ 2 \\ \hline \$16.80 \end{array}$$

(3) Ruth's weekly salary is
$$\begin{array}{r} \$196.00 \\ +\ \ 16.80 \\ \hline \$212.80 \end{array}$$

1. Round off to the nearest cent.

   a. $1\frac{1}{2} \times \$3.40$ **f.** $\$3.68 \times 1.5$

   b. $1.5 \times \$4.10$ **g.** $1\frac{1}{2} \times \$4.04$

   c. $\$3.65 \times 1\frac{1}{2}$ **h.** $1\frac{1}{2} \times \$3.73$

   d. $\$4.25 \times 1.5$ **i.** $\$4.86 \times 1.5$

   e. $1\frac{1}{2} \times \$4.60$ **j.** $1\frac{1}{2} \times \$5.13$

2. Do each of the following problems:

   a. If Jim works a regular work week of 35 hours, how many hours of overtime does he earn if he works 9, 8, 7, 8, and 9 hours per day?

   b. Mary worked 44 hours last week. Her usual work week is 35 hours. How many hours of overtime did she work?

   c. Compute the time-and-a-half rate of pay for Carmen if she earns $3.80 as her regular hourly rate of pay.

   d. What is the overtime rate of pay (time-and-a-half) for John, who earns $3.90 as his regular hourly rate of pay?

3. Complete the table below. The regular work week consists of 35 hours.

S and T Supermarket

	Employee	M	T	W	Th	F	S	Total Hours	Reg. Hours	Rate of Pay per Hour	Reg. Wages	Over-time Hours	Over-time Rate of Pay	Over-time Wages	Total Wages
a.	D. Smith	4	8	5	5	7	4			$3.45					
b.	C. Southerland	8	8	8	8	8	—	40	35	3.60	$126	5	$5.40	$27	$153
c.	N. Homere	7	7	8	9	8	7			4.10					
d.	S. Cox	6	4	3	3	4	9			3.45					
e.	B. Estrada	4	8	5	5	8	4			3.70					
f.	A. Coleman	8	8	9	8	7	—			3.55					
g.	L. Bermudez	3	5	5	8	8	8			3.90					
												Total payroll:			

4. Complete the table below. Consider that each employee works a regular 35-hour week. Use the same schedule shown for each of the 5 work days.

S and T Supermarket

	Employee	In/Out	In/Out	Hours Worked Daily	Hours Worked Weekly	Hourly Rate of Pay	Reg. Wages	Over-time Hours	Over-time Rate of Pay	Over-time Wages	Total Wages
a.	L. Medina	7:30/11:30	12:30/5:00	$8\frac{1}{2}$	$42\frac{1}{2}$	$4.10	$143.50	$7\frac{1}{2}$	$6.15	$46.13	$189.63
b.	A. Sanchez	7:30/12:00	12:30/4:00			3.84					
c.	H. Schmidt	8:00/11:00	12:00/5:30			4.00					
d.	S. Nelson	8:00/11:30	12:15/5:15			3.50					
e.	L. Guzman	7:30/11:15	12:15/5:15			3.50					
f.	Y. Mercado	9:00/1:00	1:30/5:30			3.65					
g.	J. Rivas	8:30/12:15	1:00/4:45			4.25					
h.	D. Branch	7:00/12:00	12:30/4:00			5.10					
i.	A. Ramos	7:30/11:15	12:30/5:30			4.25					
j.	C. Hunter	8:00/12:00	1:00/5:15			3.40					

5. The Best-Ever Food Company pays each of 10 employees $4.25 an hour for a 35-hour week. If 8 of the employees put in 2 hours apiece overtime at time-and-a-half, what is the Best-Ever Company's gross payroll for the week?

6. Penny earns $5.30 an hour for a 35-hour week. She earns time-and-a-half for overtime during the week and double time for weekends. What are her gross earnings if she works 9 hours a day Monday through Thursday, 7 hours Friday, 8 hours Saturday, and 5 hours Sunday?

7. Bobby earns $8.50 an hour in a bank. His regular weekly gross is $340.00. How many hours is his work week?

8. Marty and Roberto each earn $4.75 an hour for a 35-hour work week. Marty works four 8-hour days and Roberto works five 7-hour days. Whose gross wages are greater? By how much?

9. Vicky earns $6.35 an hour for a 35-hour work week. She earns time-and-a-half after 35 hours . She works 8 hours on Monday, 7 hours on Tuesday, 9 hours on Wednesday, is absent on Thursday, and works 10 hours on Friday. Is she entitled to overtime? Compute her gross pay.

10. Harry earns $5.80 an hour for a 40-hour work week. He is given a choice of 2 hours a day overtime at time-and-a-half or 8 hours on Saturday at double time. If he works on Saturday, he has additional expenses of $4.20 for carfare and $5 for lunch. How should he work the overtime to make the most money?

EXTRA FOR EXPERTS

1. An apprentice painter earned $180.00 in gross wages for working her usual 30-hour week. One week she received $210 in gross wages. How many hours in overtime at time-and-a-half did she work that week?

2. If the hourly rate of pay is $4.50 for a 40-hour week with time-and-a-half pay for overtime during the week, double time for Saturdays, and triple time for holidays, how much would a person earn for working a total of $52\frac{1}{2}$ hours, $6\frac{1}{2}$ of which were on Saturday and 4 of which were on a holiday?

5-10 *Salary (Piece-Rate Plan)*

STUDY GUIDE Garment work is one industry whose employees are paid by the piece. In other words, they earn money for every part or piece of a garment they complete. A sewer may sew only collars to a blouse. She gets paid for the number of collars she sews. On an assembly line each worker is assigned a job or special function.

The rate of pay under the *piece-rate* plan is set by union contract. Employers expect employees to produce more if they are paid by the piece.

FACT FINDING

1. One industry in which employees are paid by the piece rate plan is

    _____ _____ .

2. The "piece rate" is usually set by _____ contract.

3. Employees who work on a piece rate usually produce _____ units.

MODEL PROBLEM

Joe Torre is a newspaper delivery boy who receives $0.20 a week for each subscriber serviced.

a. How much does Joe earn each week if his route includes 125 subscribers?

b. How much does he earn at the end of a year? Joe does not pay taxes on his salary.

Solution: a. Joe's earnings per week will be

$$\begin{array}{r} 152 \\ \times\ \$0.20 \\ \hline \$30.40 \end{array}$$

b. Since there are 52 weeks in a year, he will earn

$$\begin{array}{r} \$30.40 \\ \times\ 52 \\ \hline 60\ 80 \\ 1520\ 0 \\ \hline \$1580.80 \text{ a year} \end{array}$$

1. Marilyn works in a factory as a machine operator. She receives $0.65 for each collar she sews on a blouse. If she sews 72 collars in one day, how much does she earn?

2. Jocelyn gets paid $0.75 for each collar she sews on a blouse. She worked the month of March and sewed 60 collars per day in a 22-day work month. How much pay did she gross?

3. Frank delivers newspapers to 130 subscribers.

    a. How much does he receive if he earns $0.15 per week for each subscriber?

    b. If he receives tips of $0.25 from each of 58 subscribers, $0.15 from each of 35 subscribers, and $0.10 from each of 38 subscribers, how much does he earn in tips?

4. Jack installs bumpers on new cars. He gets paid $5.75 per bumper. If he installs 8 bumpers Monday, 7 bumpers Tuesday, 10 bumpers Wednesday, 7 bumpers Thursday, and 9 bumpers Friday, what will he gross for the week?

5. Complete the payroll for the Acme Sewing Company.

	Employee	Wages per Piece	Total No. of Pieces	Daily Wages	Weekly Wages
a.	Cheryl Lilly	$0.50	96	$48.00	$240.00
b.	Rosa Genao	0.65	74		
c.	Carol Benjamin	0.72	58		
d.	Raquel Tolin	1.10	45		
e.	Yolanda Ramos	1.25	38		
f.	Carmen Lee	0.38	130		
g.	Loretta Lewis	0.17	250		
			Total payroll:		

6. Sam works in a shop as a drill press operator. He earns $4.57 an hour plus $0.15 for each additional piece he completes beyond 120 pieces per hour. If he works an 8-hour day, how much will he earn if he completes the following number of pieces for each hour he works?

	Hour	Number of Pieces Completed	Additional Pieces Completed	Additional Money Earned
a.	0–1	140	20	$3.00
b.	1–2	125		
c.	2–3	110		
d.	3–4	115		
e.	4–5	Lunch		
f.	5–6	138		
g.	6–7	124		
h.	7–8	121		
			Total additional money: Regular pay:	
			Gross wages:	

7. Marie sews monograms on men's shirts and earns $0.75 for each initial she sews. How much will she earn if her daily schedule is as follows?

	Customer	Number of Shirts	Number of Initials	Earnings
a.	James E. Blenhem	3	4	
b.	John Bennett	2	1 doz.	
c.	Lester Giddings	2	$\frac{1}{2}$ doz.	
d.	Neal L. Smith	3	11	
			Total:	

8. Cecilia works for a book company that calculates commissions on the basis of the number of books sold. She receives $0.50 per book for the first 100 sold, and $1.00 for each book sold between 101 and 200 books. One week she sells 187 books. What is her gross pay?

1. Thomas works very rapidly. Given a choice of a fixed salary for repairing cars and a minimum rate of $15.00 per car for repair work, he chose the piece rate. If he does repairs on a minimum of 4 cars per day, and was offered a fixed gross salary of $250, did he make the correct choice? Why?

2. Maria gets a regular rate of $4.70 an hour if she produces 100 units per hour. If she produces 10 extra units, she earns $0.40 additional. She worked from 8:30 till 12:30 and produced 110 units every hour. After lunch she worked from 1:30 P.M. till 4:30 P.M. and produced 100 units each hour.

    a. What were her wages for the day?

    b. If she worked at the same rate for each day of the 5-day week, what were her gross wages?

3. The foreman of a factory gets paid on the basis of the number of units produced. He supervises 11 workers who gross $72.00 each per day, They produce 1100 units. If he hires 4 more workers who work at the same rate, how many more units will be produced? If he earns $0.25 for each unit, by how much has he increased his gross pay by hiring the additional workers?

5-11 *Installment Buying*

STUDY GUIDE   *Installment buying* is also known as *buying on time*. People use installment buying when they purchase an item that costs a large sum of money. Instead of paying cash, the buyer makes an initial payment, called a *down payment,* and then many monthly payments, called *installments,* until the item is completely paid for.

```
ABC TV STORE
1  24" TV......$240.
DOWN PAYMENT....$ 60.
AMT DUE........$180.
```

The difference between the installment cost and paying cash is called the *finance charge*.

    Finance charge = installment cost − cash payment

Buying on time means that you pay more for an item than you would if you paid cash. To find the total cost of using the installment plan, you add the down payment to the number of installments times the amount of each installment.

    Cost = down payment + amount × number of installments

FACT FINDING

1. "Buying on time" is another name for _____ buying.

2. The _____ _____ is the initial payment.

3. If you buy an item for cash, you pay _____ than if you buy on time.

4. The finance charge is found by subtracting the cash price from the _____ cost.

MODEL PROBLEM

Gail wishes to purchase an automobile that sells for $6750 on the installment plan. She agrees to pay $2000 down and to make 24 monthly payments or installments of $210 each.

a. What is the total cost of the automobile purchased in this manner?

b. What is the finance charge?

Solution: a. Gail's total cost is computed as follows:
Total cost = down payment + amount × number of installments

= $2000 + $210 × 24 $210
= $2000 + $5040   × 24
                             ─────
                              840
                              420
                             ─────
                             $5040

Total cost = $7040.

b. Finance charge = installment cost − cash price
= $7040 − $6750
= $290

EXERCISES

1. A camera marked $459.95 was bought on the installment plan. A down payment of $125 was made, and 12 monthly installments of $32.40 each were paid.

   a. What was the total cost of the camera purchased on the installment plan?

   b. What was the finance charge?

2. A color television set marked $279.95 was purchased on the installment plan. Eighteen payments of $13.75 each were made in addition to $75 as a down payment.

   a. What was the total cost of the TV purchased on the installment plan?

   b. What was the finance charge?

3. Complete the table below:

	Item	Cash Price	Down Payment	Number of Installments	Am't of Each Installment	Total Cost	Finance Charge
a.	Furniture	$995	$100	12	$85.50		
b.	Dishwasher	$259.95	$25	12	$23.90		
c.	Air conditioner	$410.95	$50	15	$27.50		
d.	Automobile	$7375	$1000	36	$179.50		
e.	Watch	$210	$25	12	$18.25		

4. Teresa bought an item for $25 down. Her payments for the next 12 months were $17.95 per month. If the finance charge was $40.45, what was the cash price of the item?

5-12 *Step Function*

STUDY GUIDE Some costs are computed on a *sliding scale*. There is a charge for the first quantity and then a reduced amount for additional quantities.

Mail rates are an example of this system. A letter mailed first-class and weighing 1 oz. or less costs $0.20, and each additional ounce or part of an ounce costs $0.17. A letter weighing $\frac{1}{2}$ oz., $\frac{3}{4}$ oz., 0.9 oz., or 1 oz. costs $0.20. A letter weighing 1.2 oz. costs $0.20 + $0.17 = $0.37.

This is an example of a *step function*. It resembles a picture of a set of stair steps without any risers. The function steadily jumps from one value to another without taking on any intermediate steps. Taxicab, phone, gas, and electric companies also compute their rates in this manner.

FACT FINDING

1. The _____ function is an example of different initial and additional charges.

2. An example of a federal agency that calculates costs using the step function is the _____ _____ .

A long-distance telephone call costs $0.95 for the first 2 minutes and $0.35 for each additional minute. What is the total cost of a 5-minute call?

Solution: The first 2 minutes cost $0.95, and there are 3 additional minutes each costing $0.35. Therefore:

$$
\begin{array}{ll}
\$0.35 & \\
\underline{\times\ 3} & \\
\$1.05 & \text{(additional 3 minutes)} \\
\underline{+\ 0.95} & \text{(first two minutes)} \\
\$2.00 & \text{total cost}
\end{array}
$$

EXERCISES

1. The first-class rate for mailing a letter at the post office is as follows:

$0.20 for the first ounce
$0.17 for each additional ounce or part of an ounce

a. What is the cost of mailing a first-class letter weighing 0.3 ounce?

b. What is the cost of mailing a first-class letter weighing 4 ounces?

c. What is the cost of mailing a first-class letter weighing 7.8 ounces?

2. The telephone company charges $9.34 as the initial monthly service and equipment charge. Telephone calls are charged for as follows:

$0.09   for each call made from 8 A.M. to 9 P.M.
$0.0585 for each call made from 9 P.M. to 11 P.M.
$0.036   for each call made from 11 P.M. to 8 A.M.

a. What is the cost of 75 calls made after 8 A.M. and before 9 P.M.?

b. What is the cost of 15 calls made after 9 P.M. and before 11 P.M.?

c. What is the cost of 14 calls made after 11 P.M. and before 8 A.M.?

d. If the charges listed above are for your phone, what is the total cost of your telephone bill this month?

3. Taxicabs charge $0.90 for the first $\frac{1}{10}$ of a mile and $0.10 for each additional $\frac{1}{10}$ of a mile.

a. What is the cost of a trip of 1 mile?

b. What is the cost of a trip of 3 miles?

c. What is the cost of a trip of $2\frac{1}{2}$ miles?

d. If the cab fare was $3.20, what was the distance traveled?

4. The Hainer family has an initial monthly service and equipment charge of $10.09 for a telephone and an extension. Compute the Hainers' monthly phone bill if their local usage was as follows:

> 43 calls between 8 A.M. and 9 P.M. @ $0.09 each
> 61 calls between 9 P.M. and 11 P.M. @ $0.0585 each
> 38 calls between 11 P.M. and 8 A.M. @ $0.036 each

5. Francisco has to mail 200 letters. First-class mail costs $0.20 for the first ounce and $0.19 for each additional ounce or part of an ounce. The weight of each piece of mail is 6.2 ounces. How much will 1 letter cost? 200 letters?

6. Speedy Air Freight charges $1.25 for the first ounce and $0.75 for each additional ounce for an overnight letter to California. What is the cost to send a package weighing 1 pound 7 ounces?

7. Susan came out of the movies late one night and wanted to take a taxicab home. She lives 12 miles from the theater. If the taxi charges $0.90 for the first $\frac{1}{10}$ of a mile and $0.10 for each additional $\frac{1}{10}$ of a mile, did she have enough money to take a cab home and give the taxi driver a $0.75 tip if all she had was a $10 bill?

**EXTRA
FOR
EXPERTS**

An envelope costs $1.20 for postage under the first-class rate. If the charge is $0.18 for the first ounce and $0.17 for each additional ounce, what is the greatest possible weight of the envelope?

5-13 *Credit Cards (the Buyer)—Optional*

STUDY GUIDE A *credit card* is a way of borrowing money from a bank. When a purchase is made, you pay for it by presenting a credit card to the merchant. The bank that issues you the card agrees to pay the merchant for the amount of your purchase, and you agree to pay the bank for lending you the money.

The bank sends you a *monthly statement*, which shows each purchase and the total amount of your purchases for the month. The total of your purchases plus any amount owed from the preceding month is called the *new balance*.

If you fail to pay the new balance within 30 days, you are charged *interest* on the unpaid balance. The interest rate is approximately equal to $1\frac{1}{2}\%$ of the unpaid balance. For example, suppose that you purchase a total of $68.53 this month and you pay $25 toward that balance. You would then have to pay an interest charge of $1\frac{1}{2}\%$ on $68.53 − $25 = $43.53 on your next bill, plus the cost of your new purchases.

If you pay your bill within 30 days, no interest charge is added to your next bill.

FACT FINDING

1. The total of all your monthly purchases plus your old balance is called the _____ _____ .

2. The rate of interest the bank charges on an unpaid balance is _____%.

3. There is no interest charge if you pay the new balance within _____ days after the billing date.

MODEL PROBLEM

Mr. Reilly owed $53.89 from his previous bill and made additional purchases of $73.25 this month. What was his new balance?

Solution: (1) There is an interest charge of $1\frac{1}{2}\%$ on $53.89. Changing $1\frac{1}{2}\%$ to a decimal gives

$$\frac{1\frac{1}{2}}{100} = \frac{1.5}{100} = 0.015.$$

Multiplying 0.015 by $53.89 results in $0.81.

$$
\begin{array}{lr}
\$53.89 & \text{(two decimal places)} \\
\times\, 0.015 & \text{(three decimal places)} \\
\hline
26945 & \\
5389 & \\
\hline
\$0.80835 & \text{(five decimal places)}
\end{array}
$$

Rounding off to the nearest cent gives $0.81.

(2) Mr. Reilly's bill reads:

Old balance	$53.89
Interest	0.81
Purchases	73.25
New balance	$127.95

1. Find $1\frac{1}{2}\%$ of each of the following:

   a. $400.00   **d.** $618.00   **g.** $865.00

   b. $275.00   **e.** $426.00   **h.** $263.00

   c. $325.00   **f.** $94.00   **i.** $561.00

2. If the first column represents the current billed charges and the second column represents the amount charged during a month, what was the interest added to the bill during the month if the interest rate is $1\frac{1}{2}\%$?

	Total Current Charges	Amount Charged That Month	Interest Added to Bill
a.	$46.53	$38.50	
b.	$173.15	$95.00	
c.	$216.50	$176.40	
d.	$348.00	$319.00	
e.	$76.50	$69.35	
f.	$168.25	$168.25	
g.	$273.75	$263.65	
h.	$98.72	$89.16	

3. Find the current total charges if interest on the carry-over amount is to be calculated at $1\frac{1}{2}\%$. The previous month's balance is:

   a. $57.00   **d.** $375.25   **g.** $416.78

   b. $163.00   **e.** $180.75   **h.** $216.95

   c. $219.00   **f.** $263.15   **i.** $139.40

4. Marilyn's old carry-over balance was $102.45.

   a. What is the interest charge at 0.015 on this amount?

   b. If she purchased an additional item costing $34.99 this month, what is her current balance?

   c. If she paid $50 toward her bill, what will her interest charge be for the following month?

   d. What will be the balance that will reappear on the bill?

5. In April, Martin purchased a pair of shoes for $6.99, a pair of slacks for $15.95, and camera equipment for $30.83.

    a. What were his total purchases for the month?

    b. If he paid $50 of the amount charged, what is his unpaid balance?

    c. If the interest charge is $1\frac{1}{2}\%$ of the unpaid balance, how much is the interest charge?

    d. What is the amount of the new balance?

6. Raquel owed $210.47 on her credit card account. The bank charged her interest at the rate of $1\frac{1}{2}\%$.

    a. What was her interest charge?

    b. What was the total of her new bill?

7. Mr. Jones owed $63.97 on his credit card account. His new bill showed a new balance of $73.57. What was his charge for interest?

8. When Mrs. Rowe's bill arrived, it showed a new balance of $167.24 and an old interest charge of $2.47. What was the amount owed from the previous bill?

 **EXTRA FOR EXPERTS**

1. Sheryl wishes to buy a dirt bike. It lists for $375, and she has $210 in her saving account. If she buys the bike with her savings and pays $50 per month for the balance, how many months will it take her to complete payments if monthly interest rates are computed at $1\frac{1}{2}\%$ of the unpaid balance?

2. Michael prices a new car. He has $2500 available, and the car he wishes to purchase costs $6500. If he will be paying $150 per month for 3 years, how much interest will he be paying on this purchase?

5-14 *Credit Cards (the Merchant)—Optional*

STUDY GUIDE
Merchants or sellers benefit when customers use credit cards. They are guaranteed payment for the purchases made even if the buyer fails to pay his or her credit card bill. They don't have to worry about checks that "bounce," and there is less cash on hand in the store in case of burglaries.

In return, there are costs that the merchant must pay for the privilege of being a participant in the credit card system. There are different ways a bank may charge a participating merchant. One method used is for the merchant to pay the bank a certain percent of his average monthly credit card sale. Here is a typical table used to determine the monthly percent cost to the merchant.

Average Monthly Credit Card Sale in Dollars, Not Exceeding:	Percent
$ 10	6 %
15	4.5
20	4.43
25	4.1
30	3.87
35	3.6
40	3.2
45	3
50	2.95
75	2.6
100	2.28
125	2
150	1.7
175	1.5

FACT FINDING

1. The merchant pays for the privilege of being a _____

    _____ _____ participant.

2. The merchant pays a percent of his _____ monthly credit card sales.

3. One of the advantages of being a credit card participant for a merchant is that he or she is guaranteed _____ for purchases made.

4. If the average monthly credit card sale is greater than $50 but not exceeding $75, the cost to the merchant is _____%.

MODEL PROBLEM

Sam Baxter's record store had many credit card sales last month. There were 3246 credit card purchases amounting to a total of $38,789.70.

a. What was the amount of the average sale?

Solution: To find the average sale, divide the total amount of sales by the number of sales:

$$\frac{\$38,789.70}{3246} = \$11.95.$$

b. What percent rate must the store pay the bank for credit card privileges?

Solution: Since $11.95 does not exceed $15, the rate is 4.5%.

c. How much money is paid to the bank?

Solution: Multiply 4.5% by $13,789.70:

$$
\begin{array}{r}
13789.70 \\
\times\ 0.045 \\
\hline
68\ 94850 \\
551\ 5880 \\
\hline
\$620.53650
\end{array}
$$

Rounding off to the nearest cent gives $620.54.

d. What is the income from the 3246 purchases less the credit card charges?

Solution: Subtract:

$$
\begin{array}{r}
\$13,789.70 \\
-\ 620.54 \\
\hline
\$13,169.16
\end{array}
$$

EXERCISES

1. The Nu-Art Supply Shop has an average monthly credit card sale amounting to $23.10. What percent must they pay for credit card privileges?

2. In May, Hardware, Inc., had 342 credit card sales totaling $8040.42.

   a. What was the average sale?

   b. What percent must be paid for credit card privileges?

   c. How much money is this?

   d. How much money is actually earned by the merchant?

3. Mary's Fashion Center had credit card sales of $34.95, $67.45, $59.95, $19.95, $26.95, $110.49, $75.99, $34.99, $23.99, $16.99, $9.99, and $29.99 during July.

    a. What was the total amount of sales?

    b. What was the number of credit card sales?

    c. What was the amount of the average sale?

    d. What percent rate would Mary have to pay for credit card privileges?

    e. How much money did she pay?

    f. What was the net amount of the credit card sales?

4. Restaurants accept major credit cards readily. The restaurant "La Plume" had credit card sales of $7532.19 for 261 customers.

    a. What was the amount of the average sale?

    b. What percent rate would La Plume pay?

    c. How much money is this?

    d. What is La Plume's remainder after paying the bank?

5. S and H Appliances usually have an average monthly credit card sale of $138.10. If one of the sales in October was $2534.14, how much money did they pay the bank on that purchase?

EXTRA FOR EXPERTS

Six credit card purchases were made in the E/C Discount Center. Five of them were as follows: $24.95, $32.95, $43.95, $19.95, and $14.95. The sixth sales slip was lost, but the average credit card sale was $23.95. What was the amount of the sixth sale?

Measuring Your Progress

1. **a.** Calculate the cost of purchasing a shirt for $7.99 and a pair of jeans for $24.99.

    b. If the bill is paid with two $20 bills, how much change will be returned to the buyer?

2. How much change did Jackie receive from a $20 bill if she made a purchase of $15.63?

3. Complete each of the following:

    a. 4 oz. = __ lb.

    c. __ pt. = $\frac{1}{2}$ gal.

    b. 24 oz. = __ qt.

    d. $1\frac{1}{2}$ lb. = __ oz.

4. Multiply:

    a. $0.89
        × 4.2

    b. Divide $1.38 by 0.34 to three decimal places.

5. Which of these three items is considered the "best" buy?

Unit Price	Retail Price
$1.18 per lb.	$0.59 ½ lb.

Brand A

or

Unit Price	Retail Price
$1.20 per lb.	15¢ 2 oz.

Brand B

or

Unit Price	Retail Price
$1.186 per lb.	89¢ 12 oz.

Brand C

6. What is the retail price of an item whose unit price is $0.944 per pound if you purchase 10 oz.?

Unit Price	Retail Price
$0.944 per lb.	? 10 oz.

7. Find the unit price per pound to *three* decimal places of a pound of peas whose retail price is $1.17 for 1.31 lb.

...ice	Retail Price
...er lb.	$1.17 1.31 lb.

8. What does a package of acorn squash weigh, to the nearest hundredth pound, if its retail price is $0.38 and its unit price is $0.49/lb?

Unit Price	Retail Price
49¢ per lb.	38¢ ? lb.

9. If 7 gallons of gasoline cost $9.94, what is the cost per gallon?

10. John spent $6.43 in a store. What was his change from a $20 bill?

11. If wallpaper costs $8.95 a roll, what is the cost of 9 rolls?

12. A bus leaves its terminal at 10:30 A.M. and arrives at its destination at 1:15 P.M. on the same day. How long does the trip take?

13. Guido, who is single and claims one exemption, worked 38 hours last week. His hourly rate of pay is $4.23. He receives time-and-a-half for hours worked above 35 hours. Compute his gross pay and his net pay, remembering that he must pay state, federal, and city taxes.

14. Sam delivers papers to 241 subscribers. If he receives $0.15 for each subscriber, what are his weekly earnings?

15. Cathy bought a minibike with a $50 down payment and 10 monthly installments of $25 each. What was the total cost of the minibike?

16. The cost of duplicating a page is as follows:

$14.00 for the first 1000 copies
$1.80 for each additional 100 copies
$0.02 for each additional copy

How much would it cost to have 1450 copies made?

17. Jamie earns $4.00 per hour at his part-time job. If he worked $6\frac{1}{2}$ hours one day, how much did he earn?

Measuring Your Vocabulary

Column II contains the meanings or descriptions of the words or phrases in Column I, which are used in this chapter.

For each number from Column I, write the letter from Column II that corresponds to the best meaning or description of the word or phrase.

Column I	Column II
1. Change	a. The amount of money returned after a cash purchase is made.
2. Retail price	
3. Overtime	b. The amount of money a person earns.
4. Piece rate	c. Shopping for the best price on an item.
5. Comparison shopping	d. Take-home pay after deductions are made from the gross pay.
6. Net pay	
7. Employee	e. The cost of an item per unit of measure.
8. Credit	f. The price that an item sells for in a store.
9. Deductions	g. A worker employed by a company.
10. Down payment	h. Taxes that are subtracted from gross pay.
11. Unit price	i. The number of hours worked times the hourly wage.
12. Salary	
13. Gross wages	j. The hours an employee works above the regular work week.
14. Installment	
15. Finance charge	k. The rate of pay for every piece of work completed by an employee.
	l. The amount of money that is the difference between the cash price and the installment price.
	m. The amount of cash used as an initial payment.
	n. The monthly payment when an item is "bought on time."
	o. Money lent to a buyer in order to purchase an item on time.

Chapter 6
Probability and Statistics

6-1 *Introduction to Statistics*

STUDY GUIDE Numerical facts are called *data*. The word "data" is plural. The singular (one fact) of "data" is "datum."

Statistics is the science of collecting, organizing, and interpreting data. Many problems can be analyzed and solved by statistics.

A person who works with data is called a *statistician*. A statistician plays an important role in science, medicine, politics, economics, and other fields. He or she knows a great deal about mathematics. A statistician uses mathematics to increase the accuracy of his or her conclusions.

FACT FINDING

1. Numerical facts are called _____.

2. Statistics is the science of _____, _____, and _____ data.

3. A person who works with data is called a _____.

4. Statisticians know a great deal about _____.

5. A statistician uses mathematics to _____ the accuracy of his or her conclusions.

Weather Data for U.S. Cities
May 20, 19—

City	High Temperature	Low Temperature	Inch of Precipitation
Albany	80	48	—
Anchorage	58	41	—
Atlanta	86	60	—
Boston	66	52	—
Chicago	78	61	.23
Denver	61	43	—
Detroit	82	60	—
Honolulu	85	72	—
Houston	86	68	—
Las Vegas	91°	60	—
Minneapolis	81°	54	.09
Philadelphia	78	56	—
Phoenix	95°	68	—
San Francisco	60	52	—
San Juan, P.R.	87	75	.02
Washington, D.C.	83	63	—

Using the table shown, answer the following questions:

a. Which city had the highest temperature?

b. Which city had the lowest temperature?

c. What cities had temperatures greater than 90°?

d. Which city had the most precipitation?

e. In which cities did it rain?

Solution: a. Phoenix (95°)

          b. Anchorage (41°)

          c. Las Vegas (91°), Phoenix (95°)

          d. Chicago (.23 in.)

          e. Chicago (.23 in.), Minneapolis (.09 in.), San Juan, PR (.02 in.)

Weather in Foreign Cities
May 22, 19—

City	Average Temperature	Skies
Berlin	56	Rain
Buenos Aires	62	Partly cloudy
Brussels	59	Rain
Cairo	86	Clear
Dublin	56	Partly cloudy
Hong Kong	82	Partly cloudy
London	54	Rain
Moscow	72	Partly cloudy
Paris	56	Rain
Rio de Janeiro	72	Clear
Rome	74	Partly cloudy
Stockholm	73	Clear
Tokyo	72	Clear
Warsaw	75	Partly cloudy

1. Using the table above, answer the following questions:

   a. Which foreign city had the highest temperature?

   b. Which foreign city had the lowest temperature?

   c. What temperature occurred most often?

   d. In which foreign cities did it rain?

Elements in Seawater

Element	Milligrams per Kilogram (Parts per Million)
Chlorine	18,980
Sodium	10,561
Magnesium	1,272
Sulfur	883
Calcium	410
Potassium	390
Bromine	66
Carbon	28
Strontium	13
Boron	4.6
Fluorine	1.4
Aluminum	0.6
Iodine	0.05
Iron	0.002–0.02
Lead	0.004
Silver	0.0003
Nickel	0.0001
Gold	0.000006

2. Using the table above, answer the following questions:

a. Which element is found in largest quantity in seawater?

b. Which element is found in least quantity in seawater?

c. What are the three elements that occur in greatest quantity in seawater?

3. Mark would like to investigate the leading causes of death in the United States in a particular year. He obtained the following information from a book giving annual statistics:

Causes of Death in the United States, 19—

Rank	Cause of Death	Number of Deaths	Death Rate per 100,000 Population
1	Heart diseases	719,840	332.1
2	Cancer	380,680	175.7
3	Stroke	182,940	86.3
4	Accidents	107,204	48.2
5	Influenza and pneumonia	48,510	24.8
6	Diabetes	33,970	17.3
7	Cirrhosis of the liver	30,847	15.2
8	Arteriosclerosis	28,750	13.2
9	Suicide	28,523	13.2
10	Diseases of early infancy	22,490	11.1

a. What was the leading cause of death in the United States in the year in which Mark was interested?

b. What was the fifth leading cause of death in the United States that year?

c. What was the total number of deaths caused by heart diseases that year?

d. How many people died in accidents that year?

4. The maximum amount of earnings that can count for Social Security and on which a person pays Social Security contributions is shown in the following table:

Year	Amount	Year	Amount
1937–50	$ 3,000	1974	$13,200
1951–54	3,600	1975	14,100
1955–58	4,200	1976	15,300
1959–65	4,800	1977	16,500
1966–67	6,600	1978	17,700
1968–71	7,800	1979	22,900
1972	9,000	1980	25,900
1973	10,800	1981	29,700

Using the data from the table, answer the following questions:

a. What is the maximum amount of earnings that can count for Social Security in 1980? 1981? 1936?

b. What is the increase in the maximum amount of earnings that can count for Social Security from 1937 to 1981?

c. In what year did Social Security begin?

d. Between what two consecutive years did the maximum amount of earnings that can count for Social Security increase the most?

Estimated Membership of the Principal Religions of the World

Religion	Europe	World
Total Christian	354,874,600	956,887,700
Roman Catholic	174,141,000	542,725,000
Eastern Orthodox	63,900,600	86,600,400
Protestant	116,833,000	327,562,300
Jewish	4,842,780	15,843,271
Muslim	9,370,000	680,214,900
Zoroastrian	—	245,380
Shinto	—	61,156,000
Taoist	—	32,407,700
Confucian	32,141	187,104,300
Buddhist	243,000	299,876,200
Hindu	356,810	523,273,050
Totals	369,719,331	2,757,008,501

5. Using the data from the table above, answer the following questions:

a. What are the principal Christian religions?

b. What are the principal non-Christian religions?

c. Which Christian religion has the largest membership? smallest membership?

d. Which non-Christian religion has the largest membership?

e. What is the total membership of Christian religions? non-Christian religions?

f. Which principal religion has the fewest members?

g. Which principal religion has the most members?

h. What is the total membership of the principal religions of the total world?

6-2 Organizing Data

Data are often collected in an unorganized and random manner. In order to interpret the data, they must be organized or grouped into an ordered presentation called a *frequency table*.

The data are listed numerically or alphabetically. Each item in a set of data is assigned a *tally* or vertical mark (|). To simplify counting, every fifth tally is written as a horizontal mark passing through four tallies, as in ⊮.

The tally marks are converted into a counting number. Each counting number tells how often or frequently each item occurs. The counting number is called the *frequency number*.

FACT FINDING

1. Data are often collected in an unorganized and _____ manner.

2. The data must be organized or grouped into an ordered presentation called a _____ _____ .

3. The data are listed _____ or _____ .

4. Each item in a set of data is assigned a _____ or vertical mark.

5. Every fifth tally is written as a _____ mark passing through four tallies.

6. The tally marks are converted into a counting number or _____ number.

MODEL PROBLEM

The following set of data represents the number of baskets made by players in a free-throw contest: {6, 4, 5, 7, 8, 9, 8, 7, 3, 5, 7, 6, 5, 8, 7, 7}. Organize the data into a frequency table.

Solution: (1) Organize the data in decreasing order with the largest number first.

           (2) Assign tally marks to each number of baskets.

           (3) Find the counting or frequency number for each basket.

           (4) Check by totaling the frequency numbers and the number of items in the given set of data.

(1) Number of Baskets	(2) Tally	(3) Frequency
9	|	1
8	|||	3
7	|||||	5
6	||	2
5	|||	3
4	|	1
3	|	1

(4)   16 = Sum of frequencies
16 = Number of items in set of data

EXERCISES

1. Arrange each of the following sets of data into a frequency table:

    a. {3, 8, 5, 6, 7, 8, 9, 4, 5, 7, 8, 6, 4, 8, 9}

    b. {4, 3, 6, 8, 7, 5, 8, 4, 6, 5, 6, 7, 8, 9, 6}

    c. {6, 5, 7, 4, 8, 9, 5, 6, 7, 8, 7, 4, 8, 9, 8}

    d. {6, 5, 7, 5, 8, 5, 8, 8, 6, 5, 6, 5, 6.5, 7, 7, 5, 3, 7, 5.5, 7, 6.5}

    e. {64″, 63″, 62″, 61″, 64″, 65″, 66″, 63″, 64″, 61″, 61″, 64″, 64″}

    f. {250, 250, 260, 320, 188, 180, 210, 250, 110, 180, 188, 300}

    g. {5.0, 5.1, 5.0, 5.2, 5.1, 4.9, 5.0, 4.9, 5.3, 4.8}

    h. {4.2%, 4.3%, 4.2%, 4.3%, 4.4%, 4.3%, 4.2%, 4.3%, 4.4%, 4.3%, 4.2%, 4.4%, 4.3%, 5.4%, 4.3%}

    i. {439.8, 439.9, 440, 440.3, 440.1, 439.8, 439, 440.1}

    j. {10, 9, 8, 6, 5, 4, 3, 10, 9, 8, 8, 7, 9, 10, 8, 9, 5}

2. A checker took nine bolts from different parts of a case of bolts and measured them. Using the following set of data, arrange the measurements into a frequency table: {5.0, 5.1, 5.0, 5.2, 4.9, 5.1, 4.7, 5.1, 5.0}.

3. Linda took a test to see how fast she could read. Her teacher gave her samples from different books. Linda's speeds in words per minute were as follows: {150, 155, 160, 165, 160, 155, 150, 160, 155, 160, 165, 160}. Arrange the data into a frequency table.

4. A milk tester checked the milk from one dairy. He took half-pint samples from every fifth can of milk. Then he tested each sample for percent of butter fat. He found that the percents were as follows: {4.4%, 4.2%, 4.3%, 4.1%, 4.3%, 4.4%, 4.5%, 4.2%, 4.1%}. Arrange the data into a frequency table.

5. A checker in a brick factory weighed each hundredth brick that came out of his kilns. The following set of data represented the weight in pounds for 13 bricks: {5.0, 5.2, 5.0, 5.2, 5.1, 4.8, 5.1, 5.1, 4.9, 5.2, 4.9, 5.5, 5.1}. Arrange the data into a frequency table.

6. Miss Clark marked a set of algebra tests for the students in her class. The following set of data represents the grades of her students: {91, 86, 76, 80, 70, 73, 100, 80, 70, 72, 60, 80, 82, 87, 97, 60, 70, 70, 98, 70, 66, 73, 31, 80, 75, 76, 78, 79, 90, 100, 80, 85, 41}.

   a. Organize or group the data into a frequency table, using the six intervals noted below:

Interval	Tally	Frequency
0–54		
55–64		
65–74		
75–84		
85–94		
95–100		
Total frequencies		

   b. Using the frequency table, answer the following questions:

   (1) How many students scored under 65?

   (2) How many students scored between 65 and 74?

   (3) How many students scored over 94?

   (4) Which interval of scores has the highest frequency?

   (5) If 65 is the passing mark, how many students passed the test?

7. The following set of data represents the weights of students in Mr. Cooper's gym class: {95, 110, 140, 130, 130, 170, 160, 160, 125, 150, 135, 155, 156, 161, 165, 145, 154, 155, 157}.

   a. Compose a frequency table, using intervals of equal size as listed below:

Interval	Tally	Frequency
95–109		
110–124		
125–139		
140–154		
155–169		
170–184		

   b. Using the frequency table, answer the following questions:

   (1) How many players weigh between 95 and 109 pounds?

   (2) How many players weigh more than 154 pounds?

   (3) Which interval of weights has the lowest frequency? highest frequency?

6-3 *Interpreting Data: Finding the Mode*

STUDY GUIDE
The number, score, or other item that occurs most frequently in a set of data is called the *mode*. The word "mode" is taken from the French language and means style or fashion. The mode is found by inspection. No computation is needed. It is the measure or value that occurs most often; thus it is most "stylish."

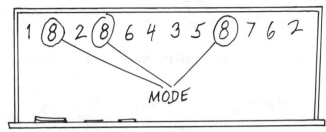

If two numbers occur most often, the data are *bimodal*. The prefix "bi-" means two.

FACT FINDING

1. The number, score, or other item that occurs most frequently in a set of data is called the _____ .

2. The word "mode" is taken from the _____ language.

3. It means style or _____ .

4. The mode is found by _____ . No computation is needed.

5. If two numbers occur most often, the data are _____ .

6. The prefix "bi-" means _____ .

MODEL PROBLEM

Using a frequency table, record the number of times each vowel (a, e, i, o, u) is used in the sentence below:

WHAT VOWEL IS THE MODE?

Solution: (1) Since the vowels are a, e, i, o, and u, set up the following frequency table:

Vowel	Tally	Frequency
a	|	1
e	|||	3
i	|	1
o	||	2
u		0

(2) Since the vowel that occurs most often is e, the mode is e.

1. After establishing a frequency table for each of the following sets of data, find the mode.

    a. {2, 3, 4, 7, 5, 7, 3, 3, 10, 4, 9}

    b. {4, 4, 21, 20, 35, 21, 27, 4}

    c. {3, 3, 1, 2, 2, 3, 2, 4, 5, 3, 4, 2, 3, 3, 4}

    d. {8, 8, 4, 3, 2, 10, 7, 8, 3, 3, 10, 11}

2. Using the frequency tables below, find the mode for each of the following sets of data:

a.

Salary	Frequency
$47,000	1
$17,000	3
$12,000	17
$11,000	4
$ 7,000	2

c.

Regents Scores for 9th-Year Math	Frequency
0–54	4
55–64	6
65–74	10
75–84	6
85–94	7
95–100	4

b.

Shoe Size	Frequency
6	3
$6\frac{1}{2}$	7
7	15
$7\frac{1}{2}$	23
8	10
$8\frac{1}{2}$	11
9	9
$9\frac{1}{2}$	3

d.

Height	Frequency
67″	7
65″	5
63″	4
62″	4
61″	3
60″	2

3. Mr. Thomas' class gave contributions to the March of Dimes fund. The table below lists the frequency of each donation. Find the contribution that was donated most often.

Contribution	Frequency
$2.00	1
0.50	10
0.25	7
0.15	11
0.10	13
0.05	3

4. Find the mode for each of the following sets of data:

    a. {4, 4, 5, 7, 7}

    b. {3, 4, 6, 6, 8, 6}

    c. {3, 2, 1, 7, 3, 7, 4, 3, 7}

5. What do the answers to Exercise 4a and 4c have in common?

6. **a.** Set up a frequency table to represent the number of vowels in this sentence:

MATHEMATICS CAN BE FUN.

    b. Using the data collected and recorded in the frequency table, find the mode.

7. **a.** Using the sentence in Exercise 6, set up a frequency table to represent the number of consonants.

    b. Using the data collected and recorded in the frequency table, find the mode. *Note:* All letters in the alphabet that are not vowels are consonants.

8. Stephen's scores for nine holes of golf are 3, 3, 3, 3, 4, 5, 6, 4, and 10. Find the mode.

9. Using the frequency table below, which shows the number of moons per planet, find the mode.

Planet	Number of Moons
Mercury	0
Venus	0
Earth	1
Mars	2
Jupiter	12
Saturn	11
Uranus	5
Neptune	2
Pluto	0

10. The ages of employees in Cury's Sport Shop are 23, 24, 19, 68, 34, 33, 24, 23, 23, 18, 35, and 34. What age appears most often?

EXTRA FOR EXPERTS

1. Why is the following set of data bimodal: {7, 7, 6, 0, 1, 10, 10, 10, 8, 8, 8, 9, 9, 10, 8}?

5

2. Charles' test scores in math are 80, 76, 92, 98, and 100. Why doesn't a mode exist?

3. The hours that Ellen spends per day writing a term paper are as follows: 1, 1, 1, 4, 5, 7, and 11.

    a. Find the mode.

    b. Is the mode a good measure for determining how many hours Ellen devoted per day to writing the paper? Why?

4. Mr. Smith wants to order ice cream to sell in his store. He examines his previous sales and finds the following data:

Flavor	Frequency (Number of Half-Gallons)
Chocolate	23
Strawberry	21
Vanilla	10

Which flavor of ice cream should he order the most of? Why?

6-4 *Interpreting Data: Finding the Mean*

STUDY GUIDE The word "average" is used often. A teacher averages grades, a motorist figures out her average mileage, and a bowler determines his bowling average.

$$(98 + 110 + 116) \div 3 = 108$$

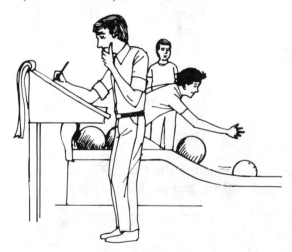

    The *arithmetic average* or *mean* is the one that people think of when they hear the word "average." To compute the arithmetic average or mean, the sum of the numbers in a set of data is divided by the number of items in the set.

    Example: In the following set of four numbers the average is 25:

$$\frac{30 + 15 + 50 + 5}{4} = \frac{100}{4} = 25$$

FACT FINDING

1. The arithmetic average or _____ is the one most people think of when they hear the word "average."

2. To compute the mean, the _____ of the numbers in a set of data is _____ by the amount of numbers.

MODEL PROBLEM

Jane wants to find the mean for the following set of test scores: {60, 80, 75, 80, 85}.

Solution: (1)   Find the sum of the test scores:

$$60 + 80 + 75 + 80 + 85 = 380.$$

(2)   Divide this sum, 380, by the number of test scores:

$$\frac{380}{5} = 76.$$

(3)   The quotient 76 is called the mean.

EXERCISES

1.   Find the mean for each of the following sets of data:

a.   {50, 48, 52}

b.   {40, 40, 40}

c.   {40, 40, 40, 50}

d.   {0, 70, 80, 90, 50}

e.   {60, 60, 65, 75, 90, 80}

f.   {60, 80, 75, 80, 85}

g.   {$4000; $8000; $15,000; $30,000}

h.   {3, 5, 8, 9, 10, 8, 8, 5}

2.   Using the frequency tables below, determine each mean:

a. Salary	Frequency		b. Contribution	Frequency
$43,000	1		$5.00	1
$20,000	2		$1.00	1
$15,000	4		$0.50	2
$12,000	7		$0.25	10
$ 7,000	1		$0.15	10
			$0.05	10

3. The heights in inches of nine professional basketball players are as follows: {66, 66, 66, 63, 71, 72, 75, 75, 78}. What is the mean height?

4. Stephen bowls in the PSAL League. In his last three-game series, he had the following set of scores: {140, 160, 180}. What was his series average (mean)?

5. Receipts from Curtis High School's football games are $212, $315, $146, $208, and $104. Find the mean value of the receipts.

6. Ann's quiz marks in chemistry are 8, 9, 7, 8, 9, 6, 10, 8, 9, 7, 8, 7, and 8. Find her mean score.

7. Barbara works on Saturdays at the supermarket. Find the mean number of hours she works per Saturday for 15 weeks, using the following set of hours: {8, 5, 6, 7, 8, 5, 4, 7, 7, 4, 8, 6, 5, 7, 8}.

8. The students in a high school science class were testing a new type of battery. They tested nine batteries for the operating life in hours and obtained the following set of data: {24, 21, 23, 24, 29, 29, 27, 28, 29}. Find the mean life of the batteries.

9. A sport shirt manufacturer has the equipment to make only one size of a man's sport shirt, and he must choose the size to make. Would he be wiser to select the mean or the mode size of the shirt bought by previous customers? Why?

10. Find to the *nearest cent* the mean for $4.32, $8.96, and $5.63.

11. Find to the *nearest dollar* the mean for $8.00, $10.00, $15.00, and $17.00.

12. Find to the *nearest thousand* the mean for 45,000; 43,000; 10,000; 93,000; and 50,000.

6-5 *Interpreting Data: Finding the Median*

STUDY GUIDE Sometimes the middle value of a given set of data gives a better representation of a set of numbers than the mean. This middle value is called the *median*. It is used in insurance for finding the average length of life expectancy, in consumer research to determine the quality of a product, in the study of drugs to determine the potency of a particular compound, and so on.

To define the median for an odd number of data, the data must be numerically ordered. The median (middle value) will be located *midway in the given set of data*.

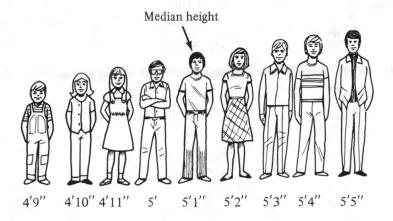

Median height

4'9" 4'10" 4'11" 5' 5'1" 5'2" 5'3" 5'4" 5'5"

To define the median for an even number of data, the data must also be numerically ordered. The median will be located *midway between the two middle values*. Therefore the median will be the average of the two middle values.

FACT FINDING

1. The middle value of a given set of data is called the _____ .

2. To determine the median value, the data must be _____

_____ .

3. For an odd number of data, the median will be located _____ in the given set of data.

4. For an even number of data, the median will be located _____

between the two _____ values.

MODEL PROBLEMS

1. Find the median for the following set of data: {20,000; 40,000; 60,000; 10,000; 8000}.

Solution: Since there are an odd number of data:

(1) Arrange the data numerically:
60,000; 40,000; 20,000; 10,000; 8000.

(2) Find the middle value:
60,000; 40,000; 20,000; 10,000; 8000.

(3) Median is 20,000.

2. Find the median for the following set of data: {20,000; 10,000; 60,000; 40,000}.

Solution: Since there are an even number of data:

    (1)   Arrange the data numerically:
        60,000; 40,000; 20,000; 10,000.

    (2)   Find the two middle values:
        60,000; <u>40,000</u>; <u>20,000</u>; 10,000.

    (3)   Find the average of the two middle values, 40,000 and 20,000.

$$\frac{40,000 + 20,000}{2} = \frac{60,000}{2} = 30,000.$$

    (4)   Median is 30,000.

EXERCISES

1. Find the median for each set of data:

   a. {6, 4, 5, 8, 7}

   b. {3, 8, 5, 6, 7}

   c. {6, 8, 6, 4, 5, 7}

   d. {10, 14, 12, 16, 18}

   e. {64″, 62″, 65″, 65″, 61″}

   f. {5, 8, 10, 12}

   g. {100, 500, 100, 550, 100, 100}

2. Using each of the following sets of data, find the median.

a. Salary (in dollars)	**c.** Dress Size
80,000	10
40,000	12
16,000	14
11,000	16
	18
b. Height (in inches)	20
	22
65	
64	
63	
62	
61	

3. Jane's test scores in geometry are 60, 70, 78, 90, and 80. Find her median score.

4. The heights (in inches) of each of the 10 members of two opposing basketball teams are as follows: 60, 62, 60, 64, 64, 65, 61, 60, 65, and 65. Find the median height.

5. The prices for 6-ounce jars of silver polish from different manufacturers are as follows: $2.00, $2.00, $2.20, $2.40, $3.60, $3.80, and $4.50. Find the median price.

6. A tire manufacturer tests 24 of his top-selling tires to determine the number of miles the tires can be driven before they wear out. The results are shown below.

Miles	Number of Tires
60,000	1
55,000	1
35,000	6
30,000	8
25,000	8

Find the median number of miles for the tires.

7. Supply the missing word in each of the following. To find the median of *N* numbers:

a. Arrange the numbers in _____ order.

b. If *N* is odd, the median is the _____ value.

c. If *N* is even, the median is the _____ value of the two middle values.

Find the median in each of the following frequency tables.

a.

Salary	Frequency
$80,000	1
$40,000	1
$16,000	10
$11,000	5

b.

Contributions (in dollars)	Frequency
1,000	2
200	2
100	2
50	3
30	4
20	6
10	8
0	3

6-6 *Simple Experiments*

STUDY GUIDE *Probability* is the science of chance. The mathematical study of games of chance began in the early part of the sixteenth century.
Today probability is used in such fields as science, insurance, and weather forecasting. The theory of probability is used to compute numerical estimates of possible outcomes of events. The chance of a particular event occurring is called the *probability of the event*.

By performing simple experiments dealing with coins, dice, cards, spinners, and other objects, it can be demonstrated that probability is a way to use numbers to predict future outcomes based on past experiences.

FACT FINDING

1. Probability is the science of _____ .

2. The theory of probability is used to compute numerical estimates of possible outcomes of _____ .

3. The chance of a particular event occurring is called the _____ of the event.

4. By performing simple _____ dealing with coins, dice, spinners, cards, and other objects, it can be demonstrated that probability is another way to use _____ to predict future outcomes based on past experiences.

MODEL PROBLEM

Mr. Jones, the baseball coach, has a free ticket to the Yankee game on Saturday. He has to choose among three deserving students: Mark, John, and David. He informs the students that he will toss two coins at the same time and give the ticket to:

Mark, if he gets a result of two heads.
John, if he gets two tails.
David, if he gets a head and a tail.

a. What are the possible outcomes of the toss?

b. What is the chance of Mark winning? John winning? David winning?

c. Which student has the best chance of winning the ticket? Why?

Solution: a. The four possibilities are HH, TT, TH, or HT, where H = heads and T = tails.

b. Mark's chance of winning are 1 out of 4 since there is only 1 chance of tossing HH out of 4 possible outcomes.
John's chance of winning are 1 out of 4 since there is only 1 chance of tossing TT out of 4 possible outcomes.
David's chance of winning are 2 out of 4 since there are 2 chances of tossing a head and a tail out of 4 possible outcomes.

c. David, since he has two ways of winning. Each of the other two boys has only one way of winning.

1. On a spinner dial with three equal regions, a pointer is spun twice. After each spin the number where the pointer stops is recorded in Table 1. After each second spin the sum is calculated. This experiment is repeated eight times.

Table 1

	Experiment							
	1	**2**	**3**	**4**	**5**	**6**	**7**	**8**
First spin	1	2	2	1	2	3	3	2
Second spin	3	3	3	3	1	3	2	2
Sum	4	5	5	4	3	6	5	4

a. If the spinner stopped at number 3, what score did the player get for the spin?

b. What is the greatest score that a player can get for one sum?

c. What is the least score that a player can get for one sum?

d. What sum occurred most often? least often?

e. What numeral do you think would be most likely to occur? least likely to occur? Why?

2. Using a spinner that is one-half red and one-half blue, a player spins twice. Each time she records where the spinner stops (on red or blue) under "First Spin" and "Second Spin" in Table 2. The experiment is repeated ten times.

Table 2

	Spins				**Results**			
	First Spin		**Second Spin**					
Exp't	**Red**	**Blue**	**Red**	**Blue**	**Red/ Red**	**One Red**	**No Red**	**At Least One Red**
1	1		1		1			1
2		1	1			1		1
3		1	1			1		1
4	1			1		1		1
5	1			1		1		1
6		1	1			1		1
7		1	1				1	
8		1	1			1		1
9		1		1			1	
10		1	1			1		1
Sum	3	7	6	4	1	7	2	8

a. Which result has the greatest number of tallies?

b. Are any two results equal?

c. Which result do you think would be most likely to occur if you tried the experiment?

d. If you repeated this experiment many times, would you expect "No Red" to have fewer tallies than "One Red"?

3. Mrs. Blake's class performed an experiment. The directions were as follows:

Take two bags, and put a green and white marble in each bag. Form three teams, one consisting of girls, one of boys, and one of the teacher. One of the girls draws a marble from each bag. If the two marbles are green, it is a point for the girls. If both are white, it is a point for the boys. If the marbles are of different colors, it is a point for the teacher. The experiment is repeated 10 times.

Using the results noted in Table 3, answer the following questions:

a. Who has the highest score?

b. If this game lasted 30 turns, which team do you think would win?

c. Fill in the "Winner" column with the word "Boys," "Girls," or "Teacher."

First Bag	Second Bag	Winner
G	G	
G	W	
W	W	
W	G	

Table 3

Exp't	First Bag	Second Bag	Girls	Boys	Teacher
			Score for:		
1	G	W			1
2	G	G	1		
3	W	W		1	
4	G	W			1
5	W	G			1
6	W	W		1	
7	W	W		1	
8	G	G	1		
9	W	G			1
10	G	W			1
Sum			2	3	5

4. Three coins are tossed at the same time. Table 4 shows the different ways they can fall. Using the results recorded in the table, answer the following questions:

    a. Which outcomes are least likely to occur?

    b. How often does HHT occur? TTH occur? HTT occur? THH occur?

Table 4

First Coin	Second Coin	Third Coin
H	H	H
H	H	T
H	T	T
H	T	H
T	T	T
T	T	H
T	H	H
T	H	T

5. A group of students performed an experiment. The directions were as follows:

Take four cards labeled A, K, Q, and J. Shuffle the cards well, and put them face down on a table. Look at the top card. After recording the kind of card, return it to the pile, shuffle, and repeat the process. Repeat this experiment for a total of 10 times.

Answer the following questions, using the results in Table 5.

    a. Number of times the J appears?

    b. Number of times the A appears?

    c. Number of times the K appears?

    d. Number of times the Q appears?

    e. Which cards appeared the most?

    f. Which cards appeared the least?

    g. If this experiment were repeated 100 times, which card(s) would you expect to show up most often? Why?

Table 5

Exp't	A	K	Q	J
1	1			
2	1			
3	1			
4		1		
5				1
6				1
7			1	
8			1	
9			1	
10		1		
Sum	3	2	3	2

If the following chart lists the total number of different ways that coins can fall, determine the total number of different ways 5 coins can fall; 6 coins can fall; 7 coins can fall.

Number of Coins	Total Number of Different Outcomes
1	2
2	4
3	8
4	16
5	?
6	?
7	?

6-7 *Computing Probabilities of Success and Failure*

STUDY GUIDE The probability of an event can be expressed as a *ratio* of two quantities:

$$\text{Probability (of success)} = \frac{\text{number of successful outcomes}}{\text{total number of possible outcomes}}$$

or

$$\text{Probability (of failure)} = \frac{\text{number of unsuccessful outcomes}}{\text{total number of possible outcomes}}.$$

For any event it is necessary to define *success* and *failure*. After this has been done, the number of successful outcomes and the number of unsuccessful outcomes are determined.

The probability of success plus the probability of failure equals 1 [$P(\text{success}) + P(\text{failure}) = 1$]. If the probability of success is given, the probability of failure is determined by subtracting the probability of success from 1, and vice versa.

Probabilities should be expressed in lowest terms.

Example: $P(\text{success}) = \frac{2}{4} = \frac{1}{2}.$

1. The probability of an event can be expressed as a _____
    of two quantities.

2. The probability of success = _____ _____

    _____ _____ divided by _____

    _____ _____ _____ _____ .

3. The probability of failure = _____ _____

    _____ _____ divided by _____

    _____ _____ _____ _____ .

4. The probability of success plus the probability of failure equals

    _____ . If the probability of success is given, the probability of
    failure is determined by subtracting the probability of success from

    _____ .

5. Probabilities should be expressed in _____ terms.

MODEL PROBLEMS

1. What are the six possible outcomes for a die (one of a pair of dice)
    to turn up when tossed once?

    Solution: 1, 2, 3, 4, 5, 6.

2. If a successful event is the tossing of an even number, find P(success)
    and P(failure).

    Solution: Since there are only three successful outcomes [2, 4, and
    6 are even] out of six possible outcomes,

$$P(\text{success}) = \frac{\text{number of successful outcomes}}{\text{total number of possible outcomes}}$$

$$= \frac{3}{6} = \frac{1}{2}.$$

$$P(\text{failure}) = 1 - P(\text{success})$$

$$= 1 - \frac{1}{2} = \frac{1}{2}.$$

    Check: $$P(\text{failure}) = \frac{\text{number of unsuccessful outcomes}}{\text{total number of possible outcomes}}$$

$$= \frac{3}{6} = \frac{1}{2}.$$

1. Supply the missing data in the last three columns of Table 1.
    Note: $P(S)$ = probability of success,
    $P(F)$ = probability of failure.

Table 1

Experiment	Outcomes	Successful Outcomes	Unsuccessful Outcomes	$P(S)$	$P(F)$
Toss a nickel	1 head, 1 tail	1 tail			
Toss a die	1, 2, 3, 4, 5, 6	1, 2, 3			
Toss 2 coins	HH, HT, TT, TH	HH			
Spin the pointer	1, 2, 3	It lands on 1.			
Roger selects a marble from the bag.	B, W, W	He selects 1 black marble.			

2. What is true about the sum of the $P(S)$ and $P(F)$ in each experiment shown in Table 1?

3. If $P(S) = \dfrac{7}{12}$, find $P(F)$.

4. If $P(F) = \dfrac{1}{3}$, find $P(S)$.

5. Supply the missing data in Table 2.

Table 2

Experiment	Success	P(success)	Failure	P(failure)
Tossing a coin	Tails	$\dfrac{1}{2}$		
Throwing a dart at a target	A hit	$\dfrac{1}{10}$		
Tossing a die	5 or 6	$\dfrac{2}{6} = \dfrac{1}{3}$		
Selecting an ace from a deck of 52 playing cards	Ace of hearts Ace of diamonds Ace of spades Ace of clubs	$\dfrac{4}{52} = \dfrac{1}{13}$		

6. Supply the missing data in Table 3.

Table 3

Sport	Success	P(success)	Failure	P(failure)
Football	Your team wins.	.800	Your team loses.	
Hockey	Your team scores a goal.	.700	Your team fails to score a goal.	
Baseball	Your team wins.	.650	Your team loses.	
Baseball	The player at bat gets a hit.	.325	The player at bat fails to get a hit.	

7. A standard deck of 52 cards is shuffled, and one card is drawn. What is the probability that the card is:

    a. The king of diamonds   **e.** A picture card (king, queen, or jack)

    b. A black card   **f.** A deuce

    c. A red card   **g.** Not an ace

    d. An ace   **h.** A red nine

8. Jane decided to guess at a question on her examination. Find the probability of getting the correct answer to the question if the question is:

    a. True-false

    b. A multiple-choice question with 5 choices

    c. A question where the choices are ''Sometimes,'' ''Always,'' or ''Never''

9. A letter is chosen at random from a given word. Find the probability that the letter is a vowel (a, e, i, o, u) if the word is:

    a. RECTANGLE   **e.** CUBE   **i.** LENGTH

    b. RHOMBUS   **f.** HYPOTENUSE   **j.** WIDTH

    c. TRIANGLE   **g.** LEG   **k.** DEPTH

    d. PARALLELOGRAM   **h.** BASE   **l.** DIMENSIONS

10. If there are 20 girls and 14 boys in a classroom, find the probability of the teacher selecting:

    a. A girl to solve a problem at the board

    b. A boy to solve a problem at the board

11. If there are 4 red blocks and 8 blue blocks in a bag, and 1 block is selected, find the probability of selecting:

    a. A red block   **b.** A blue block   **c.** A yellow block

12. If there are 4 black marbles, 4 red marbles, and 3 green marbles in a bag, and if 1 marble is taken from the bag, find the probability of selecting:

    a. 1 red marble   **d.** A marble

    b. 1 green marble   **e.** 1 purple marble

    c. 1 black marble

13. Reduce each of the following ratios to lowest terms:

a. $\frac{4}{8}$	**f.** $\frac{16}{18}$	**k.** $\frac{10}{15}$	**p.** $\frac{14}{21}$
b. $\frac{21}{42}$	**g.** $\frac{12}{16}$	**l.** $\frac{5}{15}$	**q.** $\frac{9}{27}$
c. $\frac{13}{52}$	**h.** $\frac{22}{44}$	**m.** $\frac{24}{30}$	**r.** $\frac{14}{28}$
d. $\frac{8}{16}$	**i.** $\frac{24}{32}$	**n.** $\frac{50}{75}$	**s.** $\frac{72}{90}$
e. $\frac{20}{40}$	**j.** $\frac{20}{30}$	**o.** $\frac{40}{50}$	**t.** $\frac{50}{100}$

EXTRA FOR EXPERTS

Explain why each of the following statements is incorrect:

a. Since there are 12 months in a year, the probability of Joe being born in September is $\frac{1}{12}$.

b. Since there are 7 days in a week, the probability that a person gets paid on Friday is $\frac{1}{7}$.

c. The probability of a day of the week beginning with the letter S is $\frac{2}{7}$.

6-8 Equivalent Fractions

STUDY GUIDE

There are many equivalent forms for a fraction. For example, $\frac{1}{2}, \frac{4}{8}, \frac{16}{32}, \frac{32}{64}$, etc., are *equivalent forms* of the same fraction.

The fraction $\frac{1}{2}$ is called the *reduced form* of each of these fractions.

The numerator and the denominator of the reduced form have no common factor other than 1. A *factor* is one of two or more numbers having a given product. A *prime number* is one that has only 1 and itself as factors. A *composite number* is one that is not prime.

If the numerator and the denominator of a fraction are divided by the greatest common factor, the fraction is said to be reduced to its lowest terms.

FACT FINDING

1. Fractions have _____ forms.

2. The fraction _____ is called the reduced form of the fractions $\frac{4}{8}, \frac{8}{16},$ and $\frac{16}{32}$.

3. The numerator and denominator of the reduced form have no common factor other than _____ .

4. A _____ is one of two or more numbers having a given product.

5. A _____ number is one that has only 1 and itself as factors.

6. A number that is not prime is called a _____ number.

7. If the numerator and denominator of a fraction are _____ by the greatest common factor, the fraction has been reduced to its lowest terms.

MODEL PROBLEMS

1. List the factors of each of the following numbers. Identify the number as prime or composite.

   a. 18   b. 20   c. 5   d. 11

   Solution:

Number	Factors	Prime or Composite
18	1, 2, 3, 6, 9, 18	Composite
20	1, 2, 4, 5, 10, 20	Composite
5	1, 5	Prime
11	1, 11	Prime

2. Find the greatest common factor of 16 and 32.

    Solution: Since 16 has factors of 1, 2, 4, 8, and 16, and 32 has factors of 1, 2, 4, 8, 16, and 32, the greatest common factor of 16 and 32 is 16.

3. Reduce $\frac{16}{32}$ to its simplest form.

    Solution: (1) Find the greatest common factor of 16 and 32.

    (2) Divide the numerator and the denominator by the greatest common factor.

    (3) Write the simplest form.
    Greatest common factor: 16.

$$\frac{16 \div 16}{32 \div 16} = \frac{1}{2}.$$

EXERCISES

1. In the table below, list all the factors, and state whether each number is prime or composite.

Number	Factors	Prime	or	Composite
12				
18				
25				
32				
17				
31				
28				
27				
36				
45				
54				
63				
81				
85				
90				
95				
100				

2. Find the greatest common factor for each of the following sets of numbers:

a.	12 and 16	**j.**	12 and 30	**s.**	13, 17, 41
b.	18 and 20	**k.**	25 and 100	**t.**	5, 11, 23
c.	3 and 7	**l.**	20 and 100	**u.**	10, 20, 30
d.	11 and 17	**m.**	50 and 100	**v.**	12, 24, 36
e.	25 and 24	**n.**	80 and 100	**w.**	11, 22, 33
f.	15 and 28	**o.**	80 and 90	**x.**	20, 30, 40, 50
g.	41 and 37	**p.**	75 and 100	**y.**	1, 3, 5, 7, 11
h.	42 and 52	**q.**	70 and 100	**z.**	10, 20, 80, 90
i.	15 and 30	**r.**	45 and 90		

3. Reduce each of the following fractions to its simplest form. Some fractions cannot be simplified.

a.	$\frac{12}{16}$	**f.**	$\frac{18}{40}$	**k.**	$\frac{8}{11}$	**p.**	$\frac{4}{52}$
b.	$\frac{18}{20}$	**g.**	$\frac{3}{7}$	**l.**	$\frac{22}{33}$	**q.**	$\frac{26}{52}$
c.	$\frac{8}{12}$	**h.**	$\frac{20}{30}$	**m.**	$\frac{15}{20}$	**r.**	$\frac{20}{100}$
d.	$\frac{5}{11}$	**i.**	$\frac{50}{100}$	**n.**	$\frac{6}{60}$	**s.**	$\frac{24}{72}$
e.	$\frac{20}{24}$	**j.**	$\frac{25}{100}$	**o.**	$\frac{13}{52}$	**t.**	$\frac{25}{125}$

4. Replace the "?" with the appropriate number.

a.	$\frac{1}{2} = \frac{?}{10}$	**e.**	$\frac{3}{4} = \frac{?}{16}$	**i.**	$\frac{3}{?} = \frac{9}{15}$
b.	$\frac{3}{8} = \frac{?}{16}$	**f.**	$\frac{7}{11} = \frac{56}{?}$	**j.**	$\frac{1}{?} = \frac{4}{12}$
c.	$\frac{1}{4} = \frac{?}{16}$	**g.**	$\frac{1}{8} = \frac{?}{16}$	**k.**	$\frac{1}{3} = \frac{3}{?}$
d.	$\frac{9}{10} = \frac{81}{?}$	**h.**	$\frac{2}{3} = \frac{?}{6}$	**l.**	$\frac{?}{9} = \frac{10}{18}$

Express each of the answers to Exercises 5–11 as a fraction in its simplest form.

5. If there are 60 minutes in each hour, what fractional part of an hour is 30 minutes? 45 minutes? 50 minutes?

6. A line is divided into 6 equal parts. Represent as a fraction the value given to 4 of these parts.

7. At a pizza party 15 of the 20 adults are women. What fraction of the party is men? women?

8. In the driver education class, 7 of the 21 students are girls. What fraction of the students is girls? boys?

9. Is there more gasoline in the tank of a car when the tank is one-half full or when the tank is two-fourths full? Why?

10. Two oranges are what part of a dozen?

11. Change each of the following to parts of a week:

    a. 5 days   **b.** 6 days   **c.** 7 days

12. Write the fraction $\frac{24}{60}$ in its lowest terms.

EXTRA FOR EXPERTS

Eratosthenes, a Greek mathematician, lived over 2000 years ago. He invented a "sieve" for finding prime numbers.

    a. Follow directions (1)–(4) on page 320 to create a replica of the "sieve."

    b. Using the "sieve," find all the prime numbers less than 100.

(1) Make a table of whole numbers 1 through 100 in six columns.

1	2	3	4	5	6
7	8	9	10	11	12
13	14	15	16	17	18
19	20	21	22	23	24

(2) The first number, 1, is *not* prime. Mark it out. The number 2 is the first prime number. Circle it. Then strike out all other numbers that have 2 as a factor. What is the first number greater than 2 to cross out from the table?

1̸	②	3	4̸	5	6̸
7	8̸	9	1̸0̸	11	1̸2̸

(3) Circle the first number after 2 that has not been crossed out. What prime number is it? Now cross out all factors of this prime number. Some of these numbers will have been eliminated before. What is the first number to be eliminated twice from the table? What is the first "new" number to be eliminated from the table? What is the next number larger than 3 that has not been eliminated from the table?

1̸	②	③	4̸	5	6̸
7	8̸	9̸	1̸0̸	11	1̸2̸
13	1̸4̸	1̸5̸	1̸6̸	17	1̸8̸

(4) This number, 5, is called a prime number. Follow the same process until there are no "new" numbers to be eliminated from the table. The numbers remaining (and circled) are the primes less than 100.

6-9 *Introducing Sample Spaces*

STUDY GUIDE The list or table of all possible outcomes of a probability experiment is called the *sample space*. After the sample space has been listed, the probability of a given outcome in the experiment can be easily determined by the probability ratio:

$$P(\text{success}) = \frac{\text{number of successful outcomes}}{\text{total number of possible outcomes}}.$$

1. The list or table of all possible outcomes of a probability experiment is called the _____ _____ .

2. After the sample space has been listed, the probability of a given outcome is determined by the probability ratio: P(success) = _____

_____ _____ _____ divided by

_____ _____ _____ _____

_____ .

MODEL PROBLEMS

1. A coin is tossed twice. If H stands for heads and T stands for tails, list the sample space of all possible outcomes.

Solution: The sample space is: {HH, HT, TT, TH}.

2. Using

$$P(\text{success}) = \frac{\text{number of successful outcomes}}{\text{total number of possible outcomes}},$$

express each of the following probabilities as a fraction in lowest terms:

a. P(2 heads)   c. P(at least 1 head)

b. P(1 head)   d. P(no heads)

Solution: Using the sample space in Model Problem 1,

a. $P(\text{2 heads}) = \frac{1}{4}$ since there is only *one* outcome out of *four* that results in two heads.

b. $P(\text{1 head}) = \frac{2}{4}$ since there are *two* outcomes out of *four* that result in one head.

Note: $P(\text{1 head}) = \frac{2}{4} = \frac{1}{4}$, where $\frac{1}{4}$ has been reduced to lowest terms.

c. $P(\text{at least 1 head}) = \frac{3}{4}$ since there are *three* outcomes out of *four* that result in at least one head.

d. $P(\text{no heads}) = \frac{1}{4}$ since there is only *one* outcome out of *four* that has no heads.

1. Using the sample space as shown in Model Problem 1, represent each of the following probabilities as a fraction in lowest terms:

    a. $P(2\ \text{tails})$ **c.** $P(\text{at least 1 tail})$

    b. $P(1\ \text{tail})$ **d.** $P(\text{no tails})$

2. A die is tossed.

    a. List the sample space for the experiment.

    b. Using the sample space, express each of the following probabilities as a fraction in lowest terms.

        (1) $P(\text{number 3 appears})$ (6) $P(\text{number is negative})$

        (2) $P(\text{number 7 appears})$ (7) $P(\text{number} > 3)$

        (3) $P(\text{number is less than 1})$ (8) $P(\text{number} < 5)$

        (4) $P(\text{number is even})$ (9) $P(\text{number is prime})$

        (5) $P(\text{number is odd})$ (10) $P(\text{number is composite})$

    Note: One is *not* a prime number.

3. The last digit of a telephone number can be a member of the following sample space: {0, 1, 2, 3, 4, 5, 6, 7, 8, 9}. Find the probability that the last digit is:

    a. 9   **d.** > 5 **g.** a negative number

    b. odd   **e.** < 8

    c. even   **f.** the letter Z

4. A spinner is spun twice, and the pointer lands on one of three regions of equal size, numbered 1, 2, and 3.

    a. List the sample space for the experiment.

    b. Express as a fraction in lowest terms the probability that:

        (1) Both numbers are even.   (4) Both numbers are odd.

        (2) The sum of numbers is even.   (5) The number 7 is spun.

        (3) The sum of numbers is odd.   (6) The number 1, 2, or 3 is spun.

5. The table shows all the sums for the tossing of two colored dice (one die is green, and the other die is red).

Red Die

+	1	2	3	4	5	6
1	2	3	4	5	6	7
2	3	4	5	6	7	8
3	4	5	6	7	8	9
4	5	6	7	8	9	10
5	6	7	8	9	10	11
6	7	8	9	10	11	12

(Green Die)

    a. How many ways are there of rolling a 7?

b. How many ways are there of rolling a 12?

c. What is the probability of rolling a 7?

d. What is the probability of rolling a 12?

e. What is the probability of rolling a 3? a 13?

6. a. Set up a sample space for the experiment of tossing a coin three times.

b. Using the sample space, find the probability that the three tosses of a coin will produce:

(1) 1 head   (4) 0 head

(2) 2 heads   (5) at least 1 head

(3) 3 heads   (6) at least 2 heads

7. Mrs. Smith has two children.

a. Set up a sample space for the possible sex of each child.

b. Using the sample space, find the probability that Mrs. Smith has:

(1) 2 boys   (3) at least 1 boy

(2) 2 girls   (4) at least 1 girl

8. Mona bought a "mix and match" set of clothes to take with her on vacation. The set consisted of two skirts, three blouses, and two vests. If the skirts are listed as S_1 and S_2, the blouses are listed as B_1, B_2, and B_3, and the vests are listed as V_1 and V_2, write a sample space to represent the different combinations she can dress up in during her vacation.

9. Using the sample space listed in the table, answer the following questions:

| | **Blouse** | | |
Skirt	Green (G)	Blue (B)	White (W)
green (g)	gG	gB	gW
blue (b)	bG	bB	bW
white (w)	wG	wB	wW

Note: In the table bG represents a blue skirt and a green blouse.

a. How many different outfits will have:

(1) A blue skirt and a green blouse?

(2) A green skirt and a blue blouse?

(3) Skirts and blouses of the same color?

(4) Skirts and blouses of different colors?

b. Using the responses to Exercise 9a, express each of the following probabilities as a fraction in lowest terms:

    (1) P(blue skirt and green blouse)

    (2) P(green skirt and blue blouse)

    (3) P(skirt and blouse of the same color)

    (4) P(skirt and blouse of different colors)

10. Suppose that you have a pack of well-shuffled cards labeled 2, 3, 4, 5, 6, 7, 8, and 9. Express each of the following probabilities as a fraction in lowest terms:

    a. P(draw 2)   **e.** P(picture card)

    b. P(draw even number)   **f.** P(prime number)

    c. P(draw odd number)   **g.** P(composite number)

    d. P(ace)

EXTRA FOR EXPERTS

In a game, four coins are tossed.

a.   List the sample space representing the different ways that the four coins can turn up.

b.   Using the sample space, express each of the following probabilities as a fraction in lowest terms:

    (1) P(0 head)   (5) P(4 heads)

    (2) P(1 head)   (6) P(all tails)

    (3) P(2 heads)   (7) P(at least 1 tail)

    (4) P(3 heads)   (8) P(at least 2 heads)

6-10 *Sample Spaces Involving Independent Events*

STUDY GUIDE In determining probabilities involving items that are *independent* of each other, it is often necessary to list the outcomes in a *sample space*. For example, when determining the outcomes when a coin and a die are tossed at the same time, it is helpful to list the outcomes, noting that the outcome that occurs with the coin will be completely independent of the outcome that occurs with the die. The elements of the sample space will be:

{H1, H2, H3, H4, H5, H6, T1, T2, T3, T4, T5, T6}.

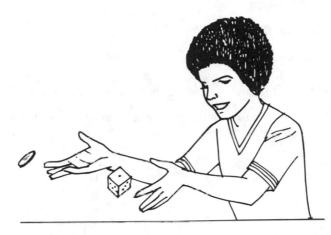

The first member of each pair represents the outcome with the coin. The second member of each pair represents the outcome with the die.

After the sample space is listed, the probability of a particular outcome can be easily determined. In addition to a probability being represented as a ratio in lowest terms, the probability may be represented as a decimal fraction or a percent.

FACT FINDING

1. Probabilities involving items that are independent are determined by first listing the _____ _____ .

2. After the sample space is listed, the probability of a particular _____ can be easily determined.

3. When a coin and a die are tossed, and the outcomes are listed, the first member of each pair represents the outcome with the _____ , and the second member of each pair represents that outcome with the _____ .

4. In addition to a probability being represented as a ratio in lowest terms, the probability may be represented as a _____ _____ or a percent.

An experiment consists of tossing a penny and a dime at the same time. Two of the members of the sample space are Ht and Hh, where the capital letter represents the face of the penny and the small letter represents the face of the dime.

a. List the sample space for the toss.

b. Using the sample space, represent each of the probabilities, $P(2$ heads$)$ and $P($at least one head$)$, as:

    (1) A ratio in lowest terms   (3) A percent

    (2) A decimal fraction

Solution: a. {Ht, Hh, Tt, Th}.

        b. $P(2$ heads$)$:

          (1) $P(2$ heads$) = \dfrac{1}{4}$ since there is only one successful outcome in the sample space out of a possible four outcomes.

          (2) $P(2$ heads$) = \dfrac{1}{4} = .25$ since $4\overline{)1.0}$.

$$\begin{array}{r} .25 \\ 4\overline{)1.0} \\ \underline{8} \\ 20 \\ \underline{20} \end{array}$$

          (3) $P(2$ heads$) = .25 = \dfrac{25}{100} = 25\%$.

        $P($at least 1 head$)$:

          (1) $P($at least one head$) = \dfrac{3}{4}$ since there are three successful outcomes out of four that result in at least one head.

          (2) $P($at least 1 head$) = \dfrac{3}{4} = .75$ since $4\overline{)3.0}$

$$\begin{array}{r} .75 \\ 4\overline{)3.0} \\ \underline{2\;8} \\ 20 \\ \underline{20} \end{array}$$

          (3) $P($at least 1 head$) = .75 = \dfrac{75}{100} = 75\%$.

1. Using the sample space of the Model Problem, represent each of the following probabilities as a fraction, a decimal fraction, and a percent:

  a. P(dime comes up tails)

  b. P(penny comes up heads)

  c. P(at least 1 tail)

  d. P(each coin lands differently)

  e. P(2 coins match)

2. Using the sample space listed in the table, represent each probability as a fraction in lowest terms, a decimal fraction, and a percent:

	A	B	C
1	1A	1B	1C
2	2A	2B	2C
3	3A	3B	3C

  a. P(odd number and letter C)

  b. P(even number and letter A)

  c. P(odd number and letter B or C)

  d. P(odd number and letter D)

3. Using the sample space listed in the table, represent each probability as a fraction in lowest terms, a decimal fraction, and a percent.

	2	4	6
4	(4, 2)	(4, 4)	(4, 6)
5	(5, 2)	(5, 4)	(5, 6)

  a. P(both numbers are the same)

  b. P(the sum of the numbers is odd)

  c. P(the sum of the numbers is even)

4. A family has three children.

    a. Write the sample space.

    b. Determine the probability that:

        (1) All are boys.

        (2) All are girls.

        (3) All are of the same sex.

        (4) The oldest is a girl.

        (5) The youngest is a boy.

        (6) Only one child is a girl.

        (7) The oldest and youngest are girls.

Represent each probability as a fraction in lowest terms, a decimal fraction, and a percent.

5. In a game, darts are thrown at two boards so that each board will contain exactly one dart. Do not count any darts that miss or land on a line. The first board contains three equal regions numbered 2, 4, and 5. The second board contains four equal regions labeled L, O, V, and E.

    a. List the sample space of all possible pairs that are outcomes for placing one dart on each board.

    b. Represent each of the following probabilities as a fraction in lowest terms, a decimal fraction, and a percent:

        (1) P(even number, V)

        (2) P(odd number, E)

        (3) P(even number, vowel)

        (4) P(odd number, consonant)

        (5) P(number 9, L)

6. The faces on a six-sided die are numbered 1 through 6. The die is rolled twice.

    a. List the sample space.

    b. Using the sample space, represent each of the following probabilities as a fraction in lowest terms, a decimal fraction, and a percent:

        (1) P(pair of even numbers)

        (2) P(pair of odd numbers)

        (3) P(sum less than 10)

        (4) P(sum of 2)

        (5) P(snake eyes) [snake eyes: 1 on each die]

        (6) P(sum of 13)

        (7) P(box cars) [box cars: 6 on each die]

EXTRA
FOR
EXPERTS

1. List the sample space when two coins are tossed; three coins; four coins; five coins.

2. Note the pattern in Extra 1. How many different outcomes would occur if six coins were tossed?

3. Two coins and a four-sided die are tossed at the same time.

    a. List the sample space.

    b. Using the sample space, represent each of the following probabilities as a fraction in lowest terms, a decimal fraction, and a percent:

        (1) *P*(all heads and even number)

        (2) *P*(all tails and odd number)

        (3) *P*(no heads and even number)

        (4) *P*(at least 1 tail and odd number)

        (5) *P*(at least 2 tails and odd number)

        (6) *P*(head and number 6)

6-11 *Predicted Number of Outcomes*

STUDY GUIDE   Experimental data give an idea of what to "predict" or to "expect" after a series of trials. To determine the number of "expected" or "predictable" successful outcomes, the probability of success is multiplied by the number of trials. This can be represented as:

Expected number of successful outcomes = probability of success × number of trials,

or

Expected number of successful outcomes = *P*(success) × number of trials.

Example: A coin is tossed four times. The desired outcome is a head. The expected number of successful outcomes

$$= \frac{1}{2} \text{ (since 1 out of 2 results is desired)} \times 4 \text{ (number of trials)}$$

$$= 2.$$

Data can be collected and used to show that the experimental data approximate the predictable outcome. However, results of individual experiments may differ from the expected or predicted number of successes.

FACT
FINDING

1. Experimental data give an idea of what to "predict" or to _____ after a series of trials.

2. To determine the number of "expected" or "predictable" successful outcomes, the probability of _____ is multiplied by the _____ of _____ .

3. Expected number of successful outcomes = $P(\text{success})$ _____ number of trials.

4. Data can be collected and used to show that the experimental data _____ the predictable outcome.

MODEL PROBLEM

If a multiple-choice test has 25 questions with five choices for each question (only one of which is correct), how many questions should a student expect to get correct if she guesses through the entire test?

Solution: (1) Find $P(\text{success})$ for each question:

$P(\text{success}) = \dfrac{1}{5}$ since *one* out of *five* choices is correct.

(2) Find the number of questions (trials): 25.

(3) Find the expected number of correct responses using the formula:

$$\begin{array}{l}\text{Expected number of successful outcomes} = P(\text{success}) \times \text{Number of trials} \\ \qquad\qquad\qquad\quad \dfrac{1}{5} \quad \times \quad\quad 25 \\ \qquad\qquad\qquad = \dfrac{25}{5} \\ \qquad\qquad\qquad = 5 \text{ correct responses.}\end{array}$$

(4) What conclusions can be drawn from the results in step 3?

Guessing will result in only 5 out of 25 correct answers, according to the expected number of outcomes.

1. Express each of the following products as a fraction in lowest terms:

a. $\frac{1}{8}$ of 40
b. $\frac{1}{2}$ of 20
c. $\frac{1}{4}$ of 100
d. $\frac{1}{8}$ of 80
e. $\frac{3}{4}$ of 16
f. $\frac{3}{4}$ of 20
g. $\frac{3}{5}$ of 50

h. $\frac{1}{2}$ of 90
i. $\frac{1}{2}$ of 16
j. $\frac{1}{5}$ of 100
k. $\frac{1}{5}$ of 25
l. $\frac{1}{5}$ of 50
m. $\frac{1}{9}$ of 90
n. $\frac{1}{11}$ of 33

o. $\frac{3}{4} \times 20$
p. $\frac{3}{4} \times 40$
q. $\frac{5}{8} \times 40$
r. $\frac{3}{8}$ of 32
s. $\frac{5}{7}$ of 42
t. $\frac{4}{9} \times 81$
u. $\frac{5}{9} \times 36$

2. Three coins are tossed at the same time. Each of the outcomes and its probability are listed in the table below. Find the predicted number of outcomes for each of the indicated number of trials.

Outcome	P(outcome)	Expected Number of Outcomes		
		40 Trials	**80 Trials**	**120 Trials**
0 Head	$\frac{1}{8}$			
1 Head	$\frac{3}{8}$			
2 Heads	$\frac{3}{8}$			
3 Heads	$\frac{1}{8}$			

3. Coin solitaire is a game played with one coin. The object of the game is for a player to toss as many times as he can before he tosses a head. On any turn a player is allowed a maximum of four tosses. Using the data listed in the table below, calculate the predicted number of outcomes for each of the indicated number of trials.

Outcome	P(outcome)	Expected Number of Outcomes		
		16 Turns	32 Turns	64 Turns
H	$\frac{1}{2}$			
TH	$\frac{1}{4}$			
TTH	$\frac{1}{8}$			
TTTH	$\frac{1}{16}$			
TTTT	$\frac{1}{16}$			

4. If a true-false test has 100 questions, how many questions should students expect to get correct, if they guess on each question?

5. If a multiple-choice test has 100 questions with four choices for each question (only one of which is correct), how many questions should students expect to get correct if they guess on each question?

6. Using the answers to Exercises 4 and 5, state which type of test (multiple-choice or true-false) is more likely to discourage guessing.

7. A spinner with three equal areas (A, B, and C) is spun 60 times. How many times will it be expected to stop on A? on B? on C?

8. What is $\frac{5}{8}$ of 64?

EXTRA FOR EXPERTS

1. If the probability of success for one trial is $\frac{2}{3}$, how many failures should be expected in 60 trials?

2. If the probability of success for one trial is $\frac{3}{8}$, how many failures should be expected in 64 trials?

Measuring Your Progress

Approximate Life Spans of Animals in Captivity

Years	Animal	Years	Animal
191	Giant tortoise	31	Dolphin
134	Eastern box turtle	30	Lion
83	Elephant	28	Giraffe
80	Freshwater oyster	28	Sea lion
69	Owl	25	Tiger
56	Eagle	25	Zebra
51	Pelican	23	Domestic cat
50	Domestic horse	22	Domestic dog
46	Jackass	20	Cow
37	Chimpanzee	14	Chicken
32	Gibbon	10	Pig
31	Grizzly bear		

1. Using the table shown above, answer the following questions:

   a. Which animal has the lowest life expectancy?

   b. Which animal has the highest life expectancy?

   c. Which animals are expected to live more than 100 years?

   d. Which animals are expected to live more than 46 years?

   e. What is the maximum life span of a domestic horse? domestic cat? domestic dog?

   f. Which animal lives eight times as long as the pig?

2. a. Arrange the following set of scores in the frequency table provided below: {60, 60, 65, 60, 100, 60, 90, 80}.

Score	Tally	Frequency
100		
90		
80		
65		
60		

   b. Using the frequency table, answer each of the following:

      (1) What is the mean score?

      (2) What is the mode?

      (3) What is the median score?

      (4) Which of the three values found in (1), (2), and (3) is most descriptive of the test scores? Why?

3. Find the mean of 78, 82, 82, and 90.

4. What is John's median bowling score for the last five games if he scored 160, 140, 182, 189, and 151?

5. What is the median age for the following set of ages: {13, 13, 14, 16}?

6. A spinner has three equal areas labeled A, B, and C.

    a. What is the probability that the spinner will stop in the area labeled A?

    b. What is the probability that the spinner will not stop in the area labeled A?

7. If $P(\text{success}) = \dfrac{1}{4}$, find $P(\text{failure})$.

8. If a coin is tossed twice, list the sample space representing all the ways the coin can fall.

9. If $P(\text{success}) = .800$, find $P(\text{failure})$.

10. Reduce the following fraction to lowest terms: $\dfrac{20}{24}$.

11. List the factors for each of the following numbers, and indicate whether each number is prime or composite:

    a. 20   **b.** 16   **c.** 13   **d.** 31

12. Find the greatest common factor for each of the following sets of numbers:

    a. 10, 20, 30   **b.** 12, 24, 36, 48

13. Replace the "?" with the appropriate number:

    a. $\dfrac{1}{6} = \dfrac{?}{12}$ **b.** $\dfrac{3}{?} = \dfrac{9}{12}$ **c.** $\dfrac{8}{12} = \dfrac{2}{?}$

14. Using the following sample space: {HT, HH, TT, TH}, find $P(\text{tossing } 2 \text{ heads})$ and $P(\text{tossing at least 1 tail})$.

15.. Express the probability of drawing an ace from a standard deck of cards as a fraction in lowest terms.

16. Using one die, express the probability of tossing an even number as:

    a. A fraction in lowest terms   **b.** A decimal fraction   **c.** A percent

17. What is $\dfrac{1}{4}$ of 100?

18. Which is a prime factor of 30?

    a. 6   **b.** 15   **c.** 1   **d.** 3

19. If $P(\text{success})$ is $\dfrac{2}{3}$, find the expected number of successful outcomes after 30 trials.

20. Express the product of the following in lowest terms: $\dfrac{3}{4} \times 100$.

Measuring Your Vocabulary

Column II contains the meanings or descriptions of the words or phrases in Column I, which are used in this chapter.

For each number from Column I, write the letter from Column II that corresponds to the best meaning or description of the word or phrase.

Column I	**Column II**
1. Data	a. Arithmetic average of a set of data.
2. Statistics	b. An ordered presentation of a set of data.
3. Statistician	c. Middle value of a set of data.
4. Frequency table	d. Ratio of the number of successful outcomes to the total number of possible outcomes.
5. Tally	e. Numerical facts.
6. "Bi-"	f. Science of collecting, organizing, and interpreting data.
7. Mode	g. Expected number of successful outcomes.
8. Mean	h. A vertical mark ($|$) indicating each item in a set of data.
9. Median	i. Prefix meaning two.
10. Probability	j. The part of a fraction above the fraction line.
11. P(success)	k. One of two or more numbers having a given product.
12. Factor	l. A number that has only 1 and itself as factors.
13. Prime number	m. The part of the fraction below the fraction line.
14. Sample space	n. The number, score, or item that occurs most frequently in a set of data.
15. Composite number	o. The result obtained in multiplication.
16. P(success) × number of trials	p. The counting number that tells how often an item in a set of data occurs.
17. French word for style or fashion	q. A set of data in which two numbers occur most often.
18. Greatest common factor	r. Number of parts out of a hundred.
19. Denominator	s. Science of chance.
20. Product	t. A person who works with data.
21. Prediction	u. A list or table of all possible outcomes of a probability experiment.
22. Numerator	v. Expectation.
23. Frequency number	w. A number that is not prime.
24. Bimodal	x. Mode.
25. Percent	y. The largest integer that is a factor of two or more integers.

Model Competency Test 1

Answer all 20 questions in this part. Write your answers on a separate answer sheet.

1. Add: 268
               443
         +   36

2. Multiply: 5341
                 × 32

3. From 993 subtract 484.

4. Divide: $17\overline{)952}$

5. The graph below shows the number of new buildings under construction in a particular city. In which year was construction lowest?

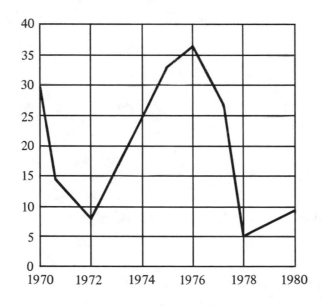

6. In three basketball games, Gary scored 20, 21, and 25 points. What was his average (mean) score for the three games?

7. Add: 2.3 + .47 + 7.68

8. Divide: 1272 ÷ 6

9. Subtract: 1.74
              −   .97

10. Multiply: $\begin{array}{r} 8.3 \\ \times\, 6 \\ \hline \end{array}$

11. Solve for x: $4x - 3 = 17$

12. Divide: $-36 \div 9$

13. The ages of five students are as follows: 14, 14, 15, 16, 17. What is the median age?

14. Solve for x: $\dfrac{3}{4} = \dfrac{x}{20}$

15. If there is a 8% sales tax, how much tax must be paid on a $21 purchase?

16. What is the sum of -3 and -7?

17. What is 20% of 70?

18. Multiply: $\dfrac{1}{3} \times \dfrac{5}{6}$

19. What is $\dfrac{2}{7}$ of 35?

20. Divide: $8 \div \dfrac{1}{2}$

PART B

Answer all 40 questions in this part. Write your answers on a separate answer sheet.

21. What number represents twenty-nine and six tenths?

(a) .296   (c) 29.6

(b) 26.06   (d) 296

22. Kris earns $3.75 per hour. How much will she earn if she works 6 hours?

(a) $20.00   (c) $22.25

(b) $20.50   (d) $22.50

23. When 596 is divided by 36, what is the remainder?

(a) 32   (c) 16

(b) 20   (d) 6

24. Linda spent $14.68 at the grocery store. How much change should she have received from a $20 bill?

    (a) $6.32   (c) $5.42

    (b) $5.32   (d) $6.68

25. Which number has the same value as $\dfrac{16}{5}$?

    (a) 11   (c) $3\dfrac{1}{5}$

    (b) $2\dfrac{1}{5}$ (d) $3\dfrac{1}{3}$

26. Mr. Rivera has $387.45 in his savings account. If he makes withdrawals of $22.50 and $18.90, how much money will be left in his account?

    (a) $34.60   (c) $346.05

    (b) $346.15   (d) $306.15

27. Jason bought a stereo on the installment plan. He paid $75 down and made 8 monthly installments of $12 each. What was the total cost of the stereo?

    (a) $96   (c) $171

    (b) $161   (d) $87

28. The graph below shows the height in inches of 50 basketball players. The greatest number of players is of which height?

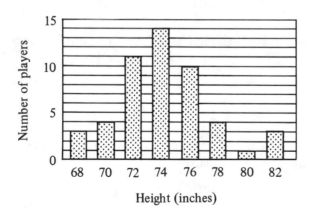

    (a) 68 in.   (c) 80 in.

    (b) 74 in.   (d) 82 in.

29. The expression 10^2 is equal to

    (a) 20   (c) 100

    (b) 200   (d) 1000

30. Mark correctly answered 16 of 20 questions on a math test. What percent of the questions did he answer correctly?

(a) 16%

(c) 12.5%

(b) 25%

(d) 80%

31. Which is a prime factor of 36?

(a) 12

(c) 6

(b) 2

(d) 9

32. Which has a value between $\frac{1}{3}$ and $\frac{3}{4}$?

(a) 1

(c) $\frac{11}{12}$

(b) $\frac{2}{3}$

(d) $\frac{4}{5}$

33. What is the least common denominator of $\frac{1}{2}$, $\frac{1}{3}$, and $\frac{2}{5}$?

(a) 6

(c) 15

(b) 10

(d) 30

34. Each side of a square measures 4.5 centimeters. What is the number of centimeters in the perimeter of the square?

(a) 15

(c) 18

(b) 20.25

(d) 9

35. Ms. Jones spent $9 for dinner. If she left a 15% tip, how much money did she leave for a tip?

(a) $.70

(c) $.90

(b) $1.35

(d) $13.50

36. On a map, 1 centimeter represents 15 kilometers. How many kilometers are represented by 2.5 centimeters on the map?

(a) 3.75

(c) 17.5

(b) 37.5

(d) 40

37. How many centimeters are equal to 3 meters?

(a) 100

(c) 1,000

(b) 300

(d) 3,000

38. A painter is paid $10 per hour. One day she begins working at 8:30 A.M. and finishes at 3:00 P.M. She does not stop for lunch. How much money should the painter be paid for the day's work?

(a) $65

(b) $60

(c) $55

(d) $50

39. If 23,684 people attended a game, what is the attendance to the nearest thousand?

(a) 20,000 (c) 24,000

(b) 23,000 (d) 30,000

40. The rectangular floor shown in the diagram below is to be covered with 1-foot-square tiles. How many tiles will be needed to completely cover the floor?

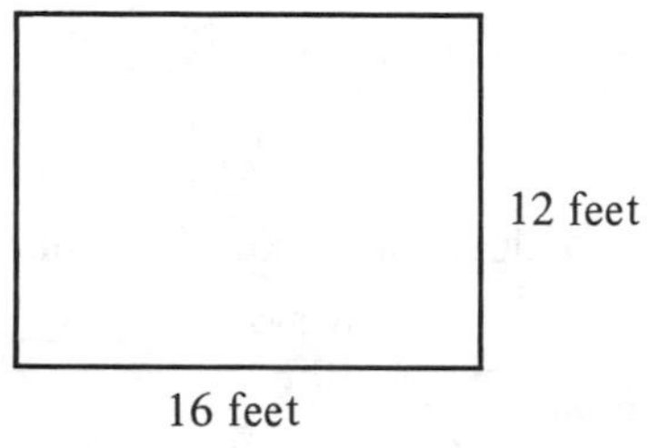

12 feet

16 feet

(a) 28 (c) 192

(b) 56 (d) 784

41. Greg bought a suit on sale for $90. If this price was 10% below the regular price of the suit, what was the regular price?

(a) $81 (c) $100

(b) $99 (d) $110

42. If 2 cans of soup cost $.39, how much will 6 cans cost?

(a) $2.34 (c) $.78

(b) $1.17 (d) $1.27

43. In its weekly budget the Dawkins family allows $55 for rent, $60 for food, $35 for utilities, and $50 for other expenses. What is the ratio of the Dawkins' food allowance to their total budget?

(a) $\frac{3}{10}$ (c) $\frac{3}{5}$

(b) $\frac{1}{2}$ (d) $\frac{1}{4}$

44. During a 5-hour period, the temperature rose from $-5°$C to $+5°$C. What was the change in temperature during that time?

(a) $-2°$C (c) $0°$C

(b) $+5°$C (d) $+10°$C

45. In the diagram below of right triangle BCD, $BD = 5$ and $CD = 4$. Using the Pythagorean theorem, $c^2 = a^2 + b^2$, determine the length of $\overline{BC}$.

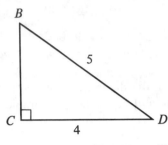

(a) 1

(b) 2

(c) 3

(d) 4

46. The circle graph below shows a budget for $300. How much money is spent for food?

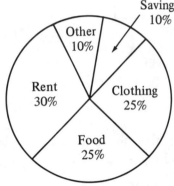

(a) $10

(b) $25

(c) $75

(d) $100

47. Which has the greatest value?

(a) 60

(b) 60%

(c) .6

(d) .06

48. On the graph below, what are the coordinates of point B?

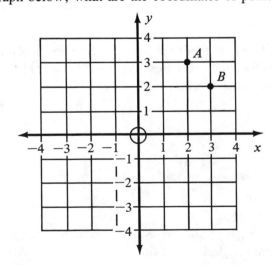

(a) (2, −3)

(b) (2, 3)

(c) (3, 2)

(d) (3, −2)

49. Wendy has 5 yards of material. If she uses only $2\frac{1}{4}$ yards, how many yards will she have left?

(a) $1\frac{1}{4}$ (c) $2\frac{1}{4}$

(b) $1\frac{3}{4}$ (d) $2\frac{3}{4}$

50. The area of a triangle can be found by using the formula $A = \frac{1}{2}bh$. What is the area of the triangle below?

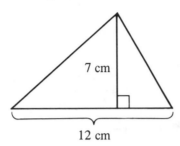

7 cm

12 cm

(a) 19 cm^2 (c) 72 cm^2

(b) 42 cm^2 (d) 144 cm^2

51. A coat was regularly priced at $102. If it was on sale for $\frac{1}{3}$ off, what was the sale price of the coat?

(a) $26   (c) $70

(b) $34   (d) $68

52. A bag contains 3 red marbles, 2 white marbles, and 1 blue marble. What is the probability of picking a red marble without looking?

(a) 1   (c) 3

(b) $\frac{1}{2}$ (d) $\frac{1}{3}$

53. If two angles of a triangle measure 30° and 85° respectively, what is the measure of the other angle?

(a) 30°   (c) 120°

(b) 65°   (d) 150°

54. What is the probability of drawing a club from a standard deck of 52 playing cards?

(a) $\frac{3}{4}$ (c) $\frac{1}{3}$

(b) $\frac{1}{2}$ (d) $\frac{1}{4}$

55. Written as a percent, the fraction $\frac{4}{10}$ is equal to

(a) .4%

(c) 14%

(b) 4%

(d) 40%

56. What is the sum of $\frac{5}{8}$ and $\frac{2}{5}$?

(a) $\frac{41}{40}$

(c) $\frac{10}{40}$

(b) $\frac{7}{40}$

(d) $\frac{31}{40}$

57. Subtract $\frac{3}{4}$ from $2\frac{1}{2}$.

(a) $2\frac{3}{4}$

(c) $1\frac{3}{4}$

(b) $2\frac{1}{4}$

(d) $1\frac{1}{4}$

58. The cost of a long-distance telephone call is as follows:

$1.12 for the first two minutes
$.48 for each additional minute

What is the total cost for a 6-minute long-distance call?

(a) $3.14

(c) $3.04

(b) $1.60

(d) $2.24

59. Using the formula $C = 2\pi r$, what is the circumference of a circle whose radius is 6? (Use $\pi = 3.14$.)

(a) 12

(c) 18.84

(b) 38.68

(d) 37.68

60. In the diagram below, what is the length of the bar?

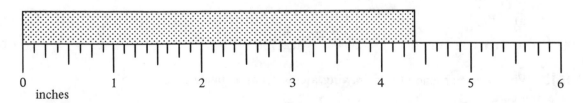

(a) 4.3 in.

(c) $4\frac{3}{4}$ in.

(b) $4\frac{3}{8}$ in.

(d) $4\frac{1}{2}$ in.

Self-Appraisal

Number of Correct Answers	Rating
57–60	Excellent
51–56	Very good
45–50	Good
39–44	Fair
33–38	Poor
0–32	Unsatisfactory

Model Competency Test 2

Answer all questions in this part. Write your answers on a separate answer sheet.

1. Add: $1 + 84 + 176 + 481$

2. Subtract: $\begin{array}{r} 3972 \\ -\ 2859 \\ \hline \end{array}$

3. Multiply: $\begin{array}{r} 411 \\ \times\ 74 \\ \hline \end{array}$

4. Divide: $15\overline{)6315}$

5. Add: $37.4 + 89.2 + 11.6$

6. Subtract: $\begin{array}{r} 38.9 \\ -\ 16.8 \\ \hline \end{array}$

7. Cara bought a blouse that cost \$11.75. She paid for it with a \$20 bill. How much change should she have received?

8. Divide: $7.2 \div 0.8$

9. A bag contains 6 red marbles, 4 blue marbles, 3 white marbles, and 2 yellow marbles. One marble is drawn without looking. Which color marble has the least chance of being drawn?

10. Multiply: -5×3

11. Multiply: $\frac{3}{4} \times \frac{1}{2}$

12. Peter wants to buy a stereo set that costs \$225. He earns \$50 each week and plans to save half of his earnings each week for the set. In how many weeks will he have saved enough money to buy the set?

13. Solve for x: $\frac{2}{3} = \frac{x}{21}$

14. Solve for x: $4x - 3 = 9$

15. Find the average (mean) of 87, 68, 75, and 78.

16. The rectangle below has a length of 5 centimeters and a width of 4 centimeters. What is the number of centimeters in the perimeter of the rectangle?

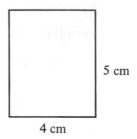

5 cm

4 cm

17. The graph below shows point P, whose coordinates are $(x, 4)$. What is the value of x?

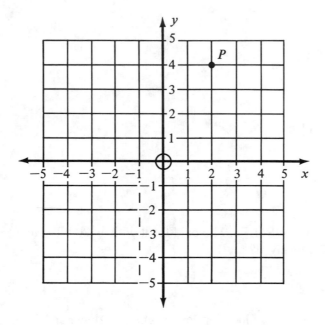

18. What is $\frac{3}{5}$ of 10?

19. If the temperature changed from $-2°C$ to $11°C$, how many degrees did the temperature rise?

20. What is a prime number between 36 and 40?

PART B

Answer all 40 questions in this part. Write your answers on a separate answer sheet.

21. Which numeral represents four thousand nine hundred twenty-four?

(a) 4900.24

(c) 49,024

(b) 4924

(d) 4,000,924

22. On a map, if 1 inch represents 16 miles, how many miles are represented by 5 inches?

(a) 64 (b) 80 (c) 21 (d) 16

23. Each week Susan has the following deductions taken from her paycheck: $6.50 for federal income tax; $1.50 for state income tax: and $1.10 for Social Security. What is the total amount deducted from Susan's check?

(a) $8.15 (b) $3.90 (c) $9.10 (d) $2.85

24. Ronny bought 3 notebooks at $.25 each and 4 pens at $.20 each. How much did he spend?

(a) $.45 (b) $1.03 (c) $1.55 (d) $1.30

25. According to the graph below, the greatest number of boys are in which height interval?

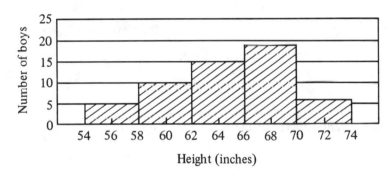

(a) 54–58 inches (b) 62–66 inches (c) 66–70 inches (d) 70–74 inches

26. The circle graph below shows the source of each dollar of government income. Where does the greatest portion come from?

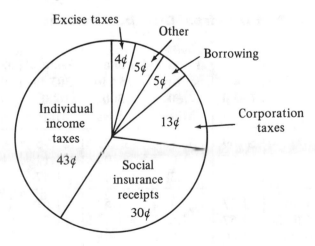

(a) Individual income taxes   (c) Corporation taxes

(b) Social insurance receipts   (d) Excise taxes

27. Joe cut 47 centimeters from a piece of wood 130 centimeters long. How long was the piece that was left?

    (a) 73 cm   (b) 83 cm   (c) 87 cm   (d) 167 cm

28. Which is equal to $\frac{14}{3}$?

    (a) $3\frac{2}{3}$ (b) $4\frac{2}{3}$ (c) $4\frac{2}{7}$ (d) 11

29. Which drawing shows perpendicular lines?

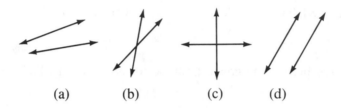

      (a)   (b)   (c)   (d)

30. Multiply: .14 × 2.2

    (a) 30.8   (c) .0308

    (b) .308   (d) .00308

31. The least common denominator of the fractions $\frac{5}{8}$ and $\frac{5}{6}$ is

    (a) 24   (c) 6

    (b) 16   (d) 10

32. According to the chart below, what is the cost of shipping a package weighing 4 lb. 7 oz. to someone who lives 400 miles from the distribution center?

Approx. Distance from Distribution Center

Shipping Weight	Local Zone	Zones 1 and 2 not over 150 Miles	Zone 3 151 to 300 Miles	Zone 4 301 to 600 Miles	Zone 5 601 to 1000 Miles
1 oz. to 8 oz.	$ .69	$ .73	$ .73	$ .75	$ .80
9 oz. to 15 oz.	1.07	1.07	1.09	1.12	1.15
1 lb. to 2 lbs.	1.38	1.38	1.42	1.47	1.53
2 lbs. 1 oz. to 3 lbs.	1.47	1.47	1.52	1.61	1.70
3 lbs. 1 oz. to 5 lbs.	1.57	1.57	1.65	1.78	1.91
5 lbs. 1 oz. to 10 lbs.	1.80	1.80	1.93	2.14	2.45
10 lbs. 1 oz. to 15 lbs.	2.15	2.17	2.39	2.77	3.22

    (a) $1.61   (b) $1.65   (c) $1.78   (d) $1.91

33. Which number is divisible by both 7 and 3?

    (a) 1   (c) 14

    (b) 9   (d) 21

34. Which has the largest value?

    (a) .16   (c) 1.15

    (b) .85   (d) 2.3

35. In the diagram below, a circle represents 100 records. What is the total number of records represented?

 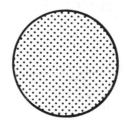 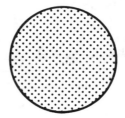

    (a) 400   (c) 275

    (b) 300   (d) 250

36. The graph below shows the amount of snowfall during the winter months. During which month does the snowfall begin to decrease?

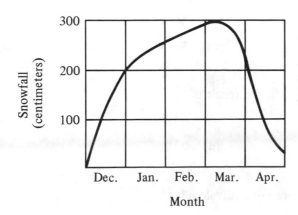

    (a) December   (c) March

    (b) January   (d) April

37. What is the approximate measure of angle Y, as shown by the protractor below?

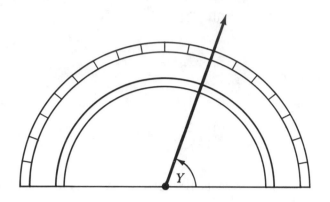

    (a) 10°   (c) 90°

    (b) 45°   (d) 70°

38. The numerical value of 5^2 is

    (a) 52   (c) 10

    (b) 25   (d) 5.2

39. The sum of $\frac{3}{5}$ and $\frac{1}{5}$ is

    (a) $\frac{4}{5}$ (c) $\frac{3}{10}$

    (b) $\frac{4}{10}$ (d) $\frac{3}{25}$

40. Jim bought a shirt for $8.00. If he had to pay 8% sales tax, what was the total cost of the shirt?

    (a) $15.00   (c) $8.64

    (b) $8.70   (d) $8.56

41. When 63.1 is divided by 10, the result is

    (a) .631   (c) 63.1

    (b) 6.31   (d) 631

42. Which pair of fractions are *not* equivalent?

    (a) $\frac{2}{4}$ and $\frac{1}{2}$ (c) $\frac{6}{9}$ and $\frac{2}{3}$

    (b) $\frac{5}{6}$ and $\frac{10}{12}$ (d) $\frac{6}{10}$ and $\frac{4}{5}$

43. Patty bought a cassette recorder for a $30 down payment and 6 monthly installments of $14 each. What was the total cost of the recorder?

(a) $264

(b) $114

(c) $144

(d) $44

44. A bus left Albany at 11:45 A.M. and reached a rest stop $2\frac{1}{2}$ hours later. What time did the bus reach the rest stop?

(a) 10:15 A.M.

(b) 12:50 A.M.

(c) 2:15 P.M.

(d) 1:15 P.M.

45. Joe is paid $4.50 an hour. How much will he earn in $6\frac{1}{2}$ hours?

(a) $29.25

(b) $26.00

(c) $24.50

(d) $24.25

46. What is 20% of 60?

(a) 12

(b) 3

(c) 6

(d) 8

47. A certain airplane has a capacity of 120 people. What is the least number of trips the airplane must make to carry a total of 500 people?

(a) $4\frac{1}{6}$

(b) 380

(c) 5

(d) 4

48. Noreen had $63.80 in her checking account. If she makes a deposit of $47.00, what will her new balance be?

(a) $7.60

(b) $100.80

(c) $110.80

(d) $16.80

49. In the circle below, which line segment is a radius?

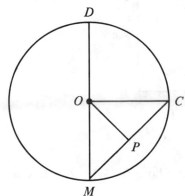

(a) $\overline{DM}$

(b) $\overline{MC}$

(c) $\overline{OM}$

(d) $\overline{OP}$

50. What percent of 10 is 5?

   (a) 5%   (c) 20%

   (b) 2%   (d) 50%

51. A recipe calls for $\frac{1}{2}$ cup of sugar. To make one-half of the recipe, how much sugar should be used?

   (a) 1 cup   (c) $\frac{1}{2}$ cup

   (b) $\frac{3}{4}$ cup   (d) $\frac{1}{4}$ cup

52. A restaurant uses 120 eggs each day. If eggs cost $.75 per dozen, what is the cost of the eggs for 1 day?

   (a) $9.00   (c) $7.50

   (b) $90.00   (d) $75.00

53. In the figure below, what is the ratio of DC to AB?

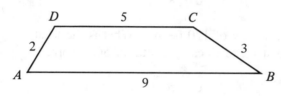

   (a) $\frac{2}{3}$ (c) $\frac{5}{9}$

   (b) $\frac{3}{2}$ (d) $\frac{9}{5}$

54. What is the area of the rectangle below?

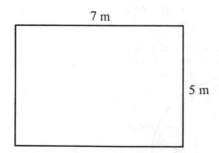

   (a) 24 m²   (c) 35 m²

   (b) 12 m²   (d) 70 m²

55. On a trip, Judy averaged 55 miles per hour for 7 hours. How many miles did she travel?

   (a) 325   (c) 305

   (b) 315   (d) 385

56. A coat is regularly priced at $95. If there is a "20% off" tag on the coat, what is the sale price of the coat?

(a) $19

(c) $66

(b) $64

(d) $76

57. What is 4.878 rounded to the nearest hundredth?

(a) 4.8

(c) 4.87

(b) 4.88

(d) 4.9

58. George Ames borrowed $900 at 10% interest from his credit union for 1 year. How much interest must he pay the credit union at the end of the year?

(a) $910.00

(c) $9.00

(b) $19.00

(d) $90.00

59. If floor tile costs $9.90 per square meter, what is the cost of covering a floor 5 meters long and 5 meters wide?

(a) $198.00

(c) $247.50

(b) $99.00

(d) $375.00

60. A square root of 16 is

(a) 1

(c) 4

(b) 16

(d) 36

Self-Appraisal

Number of Correct Answers	Rating
57–60	Excellent
51–56	Very good
45–50	Good
39–44	Fair
33–38	Poor
0–32	Unsatisfactory

Model Competency Test 3

Answer all 20 questions in this part. Write your answers on a separate answer sheet.

1. Add: 5782
          632
       + 734

2. Subtract: 578 from 2694

3. Multiply: 284
           × 32

4. Divide: 23)‾7834‾

5. Multiply: $\frac{2}{3} \times \frac{4}{5}$

6. A spinner has four equal areas labeled Red, Blue, Green, Yellow. What is the probability that the spinner will stop on Blue in the next spin?

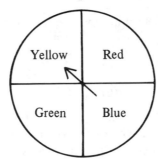

7. Multiply 12 by −4.

8. Solve for x: $\frac{2}{3} = \frac{x}{15}$

9. What is $\frac{5}{7}$ of 21?

10. What is 20% of 60?

11. Add: 8.31 + 9.2

12. Subtract: 84.71
            − 8.32

13. Multiply: 4.7
                 × 5.3

14. Divide: 9)21.6

15. Solve for x: $3x - 2 = 13$

16. Find the average (mean) of 85, 96, 55, and 76.

17. Write the numeral for three thousand sixty-five.

18. Add −5 and −9.

19. In the last 5 years, a baseball team won 83, 89, 97, 100, and 101 games. What is the median number of games won?

20. A man buys a car for $7500. The sales tax is 8% of the price of the car. How much sales tax, in dollars, must he pay?

PART B

Answer all 40 questions in this part. Write your answers on a separate answer sheet.

21. Using the formula $A = \pi r^2$, determine the area of a circle whose radius is 20. (Use $\pi = 3.14$.)

    (a) 12.56   (c) 314

    (b) 62.8   (d) 1256

22. A box contains 2 blue balls and 1 white ball. Without looking, a girl picks 1 ball from the box. What is the probability that she picks a blue ball?

    (a) $\dfrac{1}{1}$ (c) $\dfrac{1}{3}$

    (b) $\dfrac{1}{2}$ (d) $\dfrac{2}{3}$

23. What is the length of the line segment joining points A and B on the graph below?

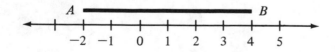

    (a) 5   (c) 2

    (b) 6   (d) 7

24. On the graph below, what are the coordinates of point A?

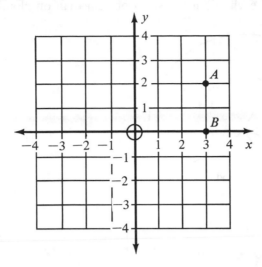

(a) (2, 2)   (c) (3, 0)

(b) (2, 3)   (d) (3, 2)

25. Each apple below represents 100,000 apples grown. What is the total number of apples represented?

(a) 400,000   (c) 500,000

(b) 550,000   (d) 350,000

26. The graph below shows Debbie's earnings as a babysitter. During which month did she earn between $20 and $25?

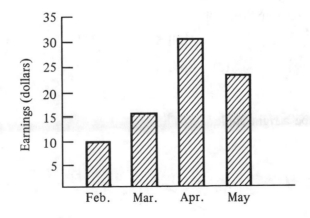

(a) February   (c) April

(b) March   (d) May

27. The circle graph below shows how Carl spends each dollar of his week's wages. How many more cents of each dollar are spent on rent than on clothing?

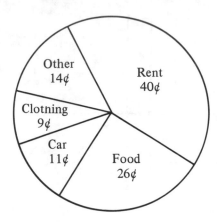

(a) 14¢

(c) 29¢

(b) 26¢

(d) 31¢

28. What is the perimeter of a triangle whose sides have lengths 4, 7, and 8?

(a) 25

(c) 19

(b) 12

(d) 6

29. As shown in the graph below, which day had the highest temperature?

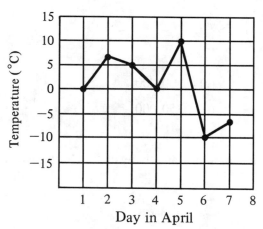

(a) April 1

(c) April 5

(b) April 2

(d) April 6

30. Which angle appears to be a right angle?

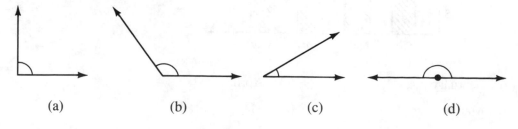

(a)   (b)   (c)   (d)

31. If a triangle has angles that measure 90° and 60°, how many degrees are in its third angle?

(a) 30°

(b) 40°

(c) 120°

(d) 150°

32. What is the area of a rectangle whose length measures 6 meters and whose width measures 4 meters?

(a) 24 m²

(b) 26 m²

(c) 18 m²

(d) 10 m²

33. Using the Pythagorean theorem, $c^2 = a^2 + b^2$, determine the length of the hypotenuse of a right triangle whose legs are 3 and 4.

(a) 5

(b) 7

(c) 12

(d) 25

34. If O is the center of the circle below, what is the line segment $\overline{OA}$ called?

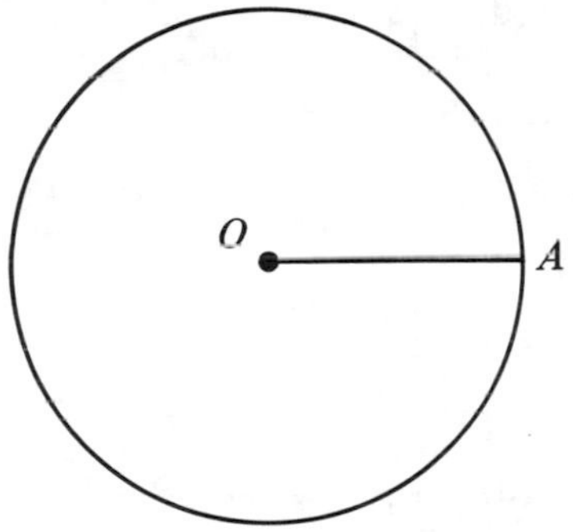

(a) an arc

(b) a chord

(c) a radius

(d) a diameter

35. Which is equal to 80%?

(a) .08

(b) .80

(c) 8.0

(d) 80

36. Written as a percent, the fraction $\frac{3}{5}$ is

(a) 2%

(b) 60%

(c) 30%

(d) 40%

37. The expression 75% is the ratio of 75 to what number?

(a) 1

(b) 17

(c) 83

(d) 100

38. The scale on a road map is 1 centimeter = 50 kilometers. If two towns are 10 centimeters apart on the map, how many kilometers apart are they in actual distance?

    (a) 250   (c) 60

    (b) 50   (d) 500

39. Which fraction has the largest value?

    (a) $\dfrac{2}{3}$ (c) $\dfrac{2}{3}$

    (b) $\dfrac{2}{5}$ (d) $\dfrac{2}{9}$

40. Add: $\dfrac{1}{3} + \dfrac{1}{4}$

    (a) $\dfrac{5}{12}$ (c) $\dfrac{2}{12}$

    (b) $\dfrac{2}{7}$ (d) $\dfrac{7}{12}$

41. Which is equal to $\dfrac{21}{4}$?

    (a) $5\dfrac{1}{5}$ (c) $4\dfrac{3}{4}$

    (b) $5\dfrac{1}{4}$ (d) $4\dfrac{1}{4}$

42. Which decimal has the smallest value?

    (a) .4   (c) .321

    (b) .72   (d) .5217

43. Subtract: $6\dfrac{2}{7}$

                   $- \, 2\dfrac{3}{7}$

    (a) $2\dfrac{1}{7}$ (c) $3\dfrac{1}{7}$

    (b) $2\dfrac{6}{7}$ (d) $3\dfrac{6}{7}$

44. Which is equal to $\dfrac{11}{25}$?

    (a) .11   (c) 4.11

    (b) .44   (d) 11.25

45. Which is a prime number?

(a) 14

(c) 16

(b) 15

(d) 17

46. Ellen earns $5 per hour. How much does she earn for working $6\frac{1}{2}$ hours?

(a) $32.50

(c) $30.50

(b) $32.00

(d) $11.50

47. A camera is regularly priced at $80. During a sale it was sold at 20% off the regular price. What was the sale price?

(a) $15

(c) $64

(b) $60

(d) $48

48. If 5 gallons of gasoline cost $6.45, what is the cost of 1 gallon?

(a) $.70

(c) $1.29

(b) $.76

(d) $1.90

49. To buy a $30.00 dress Maria put down $16.50. How much money does she still owe?

(a) $14

(c) $23.50

(b) $13.50

(d) $46.50

50. Mr. Cook had $4856 in his savings account. After making a withdrawal, he had $4516 in his account. How much did he withdraw?

(a) $240

(c) $212

(b) $340

(d) $292

51. On a certain day the sun rises at 7:00 A.M. and sets at 5:30 P.M. How many hours are there from sunrise to sunset?

(a) $1\frac{1}{2}$

(c) $10\frac{1}{2}$

(b) $8\frac{1}{2}$

(d) $12\frac{1}{3}$

52. Mrs. Conte bought a new washing machine with a $60 down payment and 12 monthly payments of $24 each. What was the total cost of the washing machine?

(a) $625

(c) $300

(b) $360

(d) $348

53. On an auto trip Joe averaged 53 miles per hour for 6 hours. How far did he travel?

    (a) 218 miles   (c) 318 miles

    (b) 328 miles   (d) 308 miles

54. A football team won 5 of the 20 games it played. What percent of its games did the team win?

    (a) 25%   (c) 12%

    (b) 20%   (d) 4%

55. A dinner costs $7.80. A 15% tip for the waiter is closest to

    (a) $.80   (c) $1.20

    (b) $1.70   (d) $1.60

56. The cost of duplicating a page is as follows:

    $12.00 for the first 100 copies
     .09 for each additional copy

How much will it cost to have 150 copies made?

    (a) $13.50   (c) $16.50

    (b) $15.00   (d) $12.00

57. Which is equivalent to 1.55 kilometers?

    (a) .155 m   (c) 135 m

    (b) 15.5 m   (d) 1550 m

58. The expression 10^3 is equal to

    (a) 1000   (c) 100

    (b) 300   (d) 30

59. What is the greatest common factor of 8 and 16?

    (a) 6   (c) 8

    (b) 2   (d) 4

60. Which value of x makes the following sentence true?

$$-5 + x = 0$$

    (a) -5 (c) $\dfrac{1}{5}$

    (b) 5   (d) 0

Self-Appraisal

Number of Correct Answers	Rating
57–60	Excellent
51–56	Very good
45–50	Good
39–44	Fair
33–38	Poor
0–32	Unsatisfactory

Index

Nonagon, 192
Number: composite, 316; directed, 60; line, 42, 66, 78; mixed, 44; prime, 316
Number line, 42, 66, 78, 87
Numerator, 44
Numerical expression, 14

Obtuse angle, 155
Octagon, 192
Order of operations, 10
Outcomes, 320, 329

Parallel lines, 180
Parallelogram, 181
Parentheses, 16
Pentagon, 192
Perimeter: of circle, 221; of polygon, 206; of rectangle, 207; of square, 207
Pi (π), 216
Place value, 117
Polygon, 192: angle, 198; diagonal, 189, 196; perimeter, 206; types: decagon, 192, dodecagon, 192, equilateral, 193, heptagon, 192, hexagon, 192, isosceles trapezoid, 181, nonagon, 192, octagon, 192, pentagon, 192, rectangle, 184; regular, 193, 198, rhombus, 184, square, 184, trapezoid, 181
Price: purchase, 235; retail, 246, 250, 254; unit, 241, 250, 251, 254
Prime number, 316, 358
Probability, 288, 306, 311, 320
Purchase price, 235
Pythagorean Rule, 173, 176

Quadrilateral, 180; diagonal, 189; types: isosceles trapezoid, 181, parallelogram, 181, rectangle, 184, rhombus, 184, square, 184, trapezoid, 181

Radius, 200, 215; radii, 200
Rational number: comparing, 71, 74; definition, 44; on a number line, 44; ordering, 68; subtraction, 98

Rectangle, 184
Regular polygon, 193, 198
Retail price, 246, 250, 254
Rhombus, 184
Right angle, 155
Rotation, 155

Sample space, 320, 325
Sector, 200
Signed numbers, 60; addition, 78; number line, 66
Sliding scale, 276
Step function, 276
Square, 184; of a number, 22; perfect, 174
Square root, 176; definition, 176; symbol, 176
Statistics, 288
Straight angle, 156
Symbol, 38

Tally, 294
Tax, sales, 235
Trapezoid, 181
Triangle: definition, 148; naming, 148; symbol, 148; types: acute, 167, equilateral, 148, isosceles, 148, obtuse, 167, right, 167, 170, scalene, 148; vertex, 148

Unit price, 241, 250, 254

Variable, 5
Vertex of triangle, 148

Wages: gross, 256, 263; hourly, 267; net, 256; overtime, 267; piece rate, 271
Whole numbers: addition, 2; division, 2; multiplication, 2; number line, 42; subtraction, 2

Yard, 114